Capitalism and Climate Change

Also by Max Koch

ROADS TO POST-FORDISM: Labour Markets and Social Structures in Europe

DIVERSITY, STANDARDIZATION AND SOCIAL TRANSFORMATION: Gender, Ethnicity and Inequality in Europe (*co-edited with L. McMillan and B. Peper*)

Capitalism and Climate Change

Theoretical Discussion, Historical Development and Policy Responses

Max Koch

Lund University, Sweden

First published 2012 by
PALGRAVE MACMILLAN

Palgrave Macmillan in the UK is an imprint of Macmillan Publishers Limited, registered in England, company number 785998, of Houndmills, Basingstoke, Hampshire RG21 6XS.

Palgrave Macmillan in the US is a division of St Martin's Press LLC, 175 Fifth Avenue, New York, NY 10010.

Palgrave Macmillan is the global academic imprint of the above companies and has companies and representatives throughout the world.

Palgrave® and Macmillan® are registered trademarks in the United States, the United Kingdom, Europe and other countries.

ISBN 978–0–230–27251–4

This book is printed on paper suitable for recycling and made from fully managed and sustained forest sources. Logging, pulping and manufacturing processes are expected to conform to the environmental regulations of the country of origin.

A catalogue record for this book is available from the British Library.

A catalog record for this book is available from the Library of Congress.

10 9 8 7 6 5 4 3 2 1
21 20 19 18 17 16 15 14 13 12

Transferred to Digital Printing in 2013

This book is dedicated to my wife Eileen Laurie, with my deepest gratitude for her love, companionship and solidarity. Thank you for never taking anything for granted.

Contents

Part IV The International Regulation of Climate Change or the Commodification of the Atmosphere

List of Tables

Acknowledgements

I would like to thank my colleagues from the Department of Social Work and Social Welfare at Lund University for their support since my move to Sweden in 2008 and during the writing of this book.

List of Abbreviations

AOSIS	Alliance of small island states
BP	British Petroleum
Btu	British thermal unit
C	Celsius
CCS	Carbon capture and storage
CDM	Clean development mechanism
CDOs	Collaterilised debt obligations
CERs	Certified emission reductions
CIS	Community of independent states
Cm	Centimetre
CMOs	Collaterilised mortgage obligations
COP	Conference of the parties
CO_2	Carbon dioxide
CO_2e	Carbon dioxide equivalent
DOEs	Designated operational entities
ECSC	European Community of Steel and Coal
EEC	European Economic Community
EIA	Energy Information Administration
ERUs	Emission reduction units
EU	European Union
EUAs	European Union allowances
EU ETS	European Union Emissions Trading System
FDI	Foreign direct investment
FED	Federal Reserve
GATS	General Agreement on Trade in Services
GATT	General Agreement on Tariffs and Trade
GDP	Gross domestic product
GM	General Motors
Gt	Gigatonne
G8	Group of Eight
G20	Group of Twenty
G77	Group of 77
IEA	International Energy Agency
IMF	International Monetary Fund
IPCC	International Panel on Climate Change

HCFC-22	Chlorodifluoromethane
HDI	Human development index
HFC-23	Trifluoromethane
JI	Joint implementation
Km	Kilometre
MOP	Meeting of the parties (of the Kyoto Protocol)
Mt	Million tonnes
Mtoe	Million tonnes of oil equivalent
M&As	Mergers and acquisitions
NAP	National allocation plan
NASA	National Aeronautics and Space Administration
NGO	Non-government organisation
OECD	Organisation of Economic Cooperation and Development
OPEC	Organisation of the Petroleum Exporting Countries
Ppm	Parts per million
PPP	Purchasing power parity
T	Tonne
Toe	Tonne of oil equivalent
TRIMS	Agreement on Trade-Related Investment Measures
TRIPS	Trade-Related Aspects of Intellectual Property Rights
UNCTAD	United Nations Conference on Trade and Development
UK	United Kingdom
UN	United Nations
UNDP	United Nations Development Programme
UNEP	United Nations Environmental Programme
UNFCCC	United Nations Framework Convention on Climate Change
US	United States
USSR	Union of Soviet Socialist Republics
WCPR	World Climate Research Programme
WMO	World Metereological Organisation
WRI	World Resources Institute
WTO	World Trade Organisation
WWF	World Wildlife Funds

Introduction

James Hansen (2008, pp. 7–8), a climate scientist of global renown and director of the National Aeronautics and Space Administration's (NASA) Goddard Institute for Space Studies, is unambiguous in his diagnosis of climate change and its profound implications: 'Our home planet is dangerously near a tipping point at which human-made greenhouse gases reach a level where major climate changes can proceed mostly under their own momentum.' Hansen is also outspoken on what should be done about the issue. The 'only resolution' he sees for the future of the human species is 'to move to a fundamentally different energy pathway within a decade'. Yet Hansen also identifies a 'huge gap between what is understood about global warming – by the scientific community – and what is known about global warming – by those who need to know: the public and policymakers' (Hansen, 2008, p. 11). And he is also very aware that preserving our planet as we know it 'will not be easy: special interests are resistant to change and have inordinate power in our governments [...]' (p. 8). Social scientists can help with illuminating what the United Nations Development Programme (UNDP) calls the 'gap between scientific evidence and political response' (UNDP, 2007, p. 4) and with understanding and questioning the power asymmetries that undermine the political reactions to climate change that climate scientists deem necessary. This is the main focus of this book. It deals with climate change as a *social* issue and aims to contribute towards understanding why the knowledge of a serious environmental anomaly – and this for more than a century – has not led to profound changes in the way people work, consume the fruits of their labour and live together in society. The idea for writing this book developed as a result of my dissatisfaction with the currently available literature on climate change, on the environmental crisis and on the financial and socio-economic

1

crisis of the advanced capitalist countries. While there is a growing body of literature in political economy circles on finance-driven capitalism and its recent crisis, this debate has rarely been linked to environmental issues, let alone climate change. Conversely, books on climate change often have a natural science focus, are technical in character and do not attempt to embed the issue into a theoretical framework for understanding contemporary society. I will, therefore, focus on the homology between two developments, capitalism and climate change, which has not been the focus of analysis to date. In this introduction, I will first summarise the most relevant natural science findings on climate change and then sketch out the potential contribution that social science could make towards understanding the issue.

For laypersons, including social scientists, climate change as a natural phenomenon is difficult to grasp. Thankfully, the world's leading climate scientists, united under the umbrella of the United Nation's (UN) International Panel on Climate Change (IPCC), periodically issue state-of-the-art evidence on climate change in popular forms that policymakers, social scientists and other non-experts can understand (see for example,Le Treut et al., 2007; IPCC, 2007a). Energy reaches the earth from the sun in the shape of sunlight and is absorbed and radiated back into space as infrared glow. Roughly one-third of the solar energy that reaches the top of the earth's atmosphere is reflected directly back into space, while the remaining two-thirds are absorbed by the surface and, to a lesser extent, by the atmosphere. The understanding of the greenhouse effect dates back to the early nineteenth century and the work of the French mathematician and physicist Jean-Baptiste Fourier, who examined the differential between the energy coming in from the sun and the energy going out as infrared radiation. Since the planet turned out to be much warmer than he had expected on the basis of his calculations, he concluded that something trapped the infrared radiation. He found that the atmosphere acts like a blanket, conserving a proportion of the heat that would otherwise radiate into space and thereby making the earth habitable for humans, animals and plants. This process came to be called 'greenhouse effect', because the atmosphere traps a proportion of the radiation in the same way in which the glass walls of a greenhouse reduce airflow and increase the temperature of the air inside it. Some greenhouse effect is, therefore, 'necessary to create the conditions for life' (Simms, 2005, p. 2), because without this effect average temperatures on earth would be approximately 30°C colder.

About three decades later, in the mid-nineteenth century, the Irish scientist John Tyndall studied the atmosphere and the way and to what extent different gases were capable of absorbing heat. He found that the gases that make up most of the atmosphere, nitrogen and oxygen (78 and 21 per cent, respectively), were not barriers to heat loss, while molecules that are much less common, such as water vapour, carbon dioxide (CO_2) and methane absorbed heat. This is why they are now known as 'greenhouse gases' (or trace gases). It is because these gases are present in the atmosphere in relatively small amounts – CO_2 as one of the major greenhouse gases makes up less than 0.04 per cent of the composition of the air – that additional emissions of them, caused by human economic and cultural activity, can have such a huge impact on the climate (Giddens, 2009; Stern, 2009). Since the percentage figures are so small, scientists use the calculation 'parts per million' (ppm) to measure the level of greenhouse gases in the air. One ppm is equivalent to 0.0001 per cent. Since CO_2 is the most important greenhouse gas in terms of volume, it is often used as a standard of measurement in assessing overall greenhouse gas emissions in the atmosphere (Giddens, 2009, p. 18).[1]

The Swedish chemist Svante Arrhenius is credited as the first scholar to link the burning of fossil fuels with potentially far-reaching changes in the earth's climate. Not only did Arrhenius confirm Tyndall's finding that water vapour and CO_2 prevented the occurrence of freezing temperatures that would be hostile to life, but he also found that enough new CO_2 had already been released into the atmosphere to make a significant difference to the ground temperature of the planet over time. In his 1895 paper 'On the Influence of Carbonic Acid in the Air upon Temperature of the Ground', Arrhenius predicted that the build-up of CO_2 as a result of economic activities that relied on burning fossil fuels would lead to global warming (see Simms, 2005, p. 16). He also presented calculations of the temperature increase resulting from a doubling of CO_2 concentrations in the atmosphere from mid-nineteenth-century levels of around 285 ppm. Remarkably accurately, he suggested that a 40 per cent increase or decrease in the atmospheric abundance of the greenhouse gas CO_2 would trigger glacial advances and retreats. A century later, the IPCC would confirm that 'CO_2 did indeed vary by this amount between glacial and interglacial periods' (Le Treut et al., 2007, p. 105).

Over millions of years, the carbon that was once in the earth's atmosphere was removed and stored. Carbon is itself the remains of past life, especially of the 'first wave of gigantic ferns and giant trees' (Clark and York, 2005, p. 402), and it was absorbed and contained in non-living forms such as oceans, glaciers and rocks. Up to the present day, these have served as sinks, limiting the accumulation of CO_2 in the atmosphere. The atmospheric concentration of CO_2 is also influenced by other components of the climate system, notably plants taking it 'out of the atmosphere and converting it (and water) into carbohydrates via photosynthesis' (Le Treut et al., 2007, p. 116). This process is known as 'carbon fixation'. Some carbon re-enters the atmosphere as CO_2 through the respiration of plants and animals, while another percentage of it is passed on to other species, and then onwards, through the food chain, as dead matter. Thus CO_2 is permanently released into the atmosphere, 'only to be recirculated to the earth through a variety of pathways in natural processes' (Clark and York, 2005, p. 402) – a cyclical process that, according to the world's climate scientists' consensus, has operated in a relatively constrained manner, sustaining the temperature of the earth throughout the four previous glacial and warming cycles. During all this long period, there has been a 'high correlation between atmospheric concentrations of CO_2 and temperature' (UNDP, 2007, p. 31). Yet the complex and delicate climate cycle is brought out of balance if increasing amounts of greenhouse gases such as CO_2 are added to the atmosphere. This is precisely what has happened over the last 160 years or so. According to the IPCC, human economic activities, primarily the burning of fossil fuels and the clearing of forests, have 'greatly intensified the natural greenhouse effect, causing global warming' (Le Treut et al., 2007, p. 115). Human societies are today returning CO_2 into the atmosphere around 'one million times faster than natural processes remove it' (Simms, 2005, p. 3). And we keep adding on to the concentrations or stocks of greenhouse gases in the atmosphere year by year, which is much more than the planet can absorb during the carbon cycle. Indeed, when referring to the historically unprecedented accumulation of CO_2 in the atmosphere, Clark and York (2005, p. 403) identify a 'rupture' or 'rift' in the carbon cycle caused by modern society.

Arrhenius's findings on the impact of greenhouse gases such as CO_2 on the world climate were ignored for a long time, and it was not until the 1930s that they were taken up again. Writing in 1938, G. S. Callendar resurrected Arrhenius's theory about fossil fuel-driven global warming from the greenhouse effect and found that a doubling of atmospheric CO_2 concentration would result in an increase in the mean global temperature of $2°C$, with considerably more warming at the poles (Le Treut et al., 2007, p. 105). However, Callendar was misled by the economic downturn following the Great Depression into thinking that 'increasing industrial efficiency had actually stabilised the level of production of greenhouse gases' (Simms, 2005, p. 20). It was not until after the

Second World War, in the 1950s, that the interdisciplinary field of climate science began to develop and the world started to wake up to the idea that human beings themselves could be responsible for average temperatures – something it seems unlikely we could have an influence upon. The composition and functions of the atmosphere were henceforth understood in more sophisticated ways. For example, understandings of the concept of a climate system were improved in 1958, when it was discovered that a greenhouse effect was responsible for the high temperatures on the surface of Venus. By the end of the 1960s, research was pointing to potentially catastrophic effects on the earth as a habitable planet if the Antarctic ice sheets were to collapse (Simms, 2005, pp. 18–22). The 1970s heralded the birth of the green parties and, with them, a greater general awareness of environmental issues. Critical scientists such as the Club of Rome published *The Limits to Growth* in 1972, and thermodynamic economists such as Nicolas Georgescu-Roegen and Herman Daly questioned the sustainability of Western work processes and lifestyles upon a finite earth (Chapter 1). Yet nearly another decade passed before the most authoritative body for monitoring climate change and its implications, the IPCC, was founded in 1987. Since that time, the IPCC's task has been to study climate change, to report the most relevant findings in climate science and to consult governments engaged in regulating this issue (see also Chapter 12). In the fourth and most recent of its reports on the state of scientific opinion, published in 2007, the IPCC (2007a) states that 'warming of the climate system is unequivocal'.[2] Climate change is now a scientifically established fact.[3] Most of the remaining sections of the document are presented in terms of probabilities and scenarios. There is, for example, a '90 per cent probability' that observed warming is the result of human activity through the introduction of greenhouse gases into the atmosphere and that these come from the consumption of fossil fuels in production and travel as well as from new forms of land use and agriculture. The report is unambiguous about the overwhelming body of scientific evidence that links rising temperatures to increased atmospheric concentrations of CO_2 and of other greenhouse gases. The world has now reached the warmest level on record in the current interglacial period, which began around 12,000 years ago': 'Eleven of the twelve years (1995–2006) rank among the 12 warmest years in the instrumental record of global surface temperature' since 1850 (IPCC, 2007a, p. 5). The earth has warmed by 0.7° over the past 100 years. Observations from all parts of the world show progressive increases in average air and sea temperatures (IPCC, 2007a; UNDP, 2007, p. 31).

The available evidence further indicates that 'at no time during the past 650,000 years has the CO_2 content of the air been as high as it is today' (Giddens, 2009, p. 18). Oscillating at levels below 290 ppm over many thousands of years, CO_2 stocks have increased by one-third since the mid-nineteenth century – at a growth rate 'unprecedented during at least the last 20,000 years' (UNDP, 2007, p. 31). By 2008, the CO_2 concentration in the atmosphere had reached 387 ppm, and it has continued to rise by around 2 ppm each year. Hence, what is qualitatively different and new about the current warming cycle is the rapid rate at which CO_2 concentrations are increasing. This increase is aggravated by the limited and apparently decreasing capacity of the earth's ecological systems, particularly oceans and forests, to absorb these emissions. Oceans naturally absorb just 0.1 gigatonne (Gt) more CO_2 per year than they release. Yet 'now they are soaking up an extra 2 Gt a year – more than 20 times the natural rate' (UNDP, 2007, p. 33). The result is that oceans are becoming warmer and increasingly acidic. Climate scientists point out the danger that this will further damage the oceans' ability to absorb CO_2. The absorptive capacity of forests, for their part, varies by type and location. About ten times more CO_2 is stored in native tropical forests (approx. 500 tonnes per hectare) than in northern native forests. Hence, the more tropical forests disappear, the less CO_2 is absorbed. According to Stern (2009, p. 25), the annual deforestation rates for native tropical forests between 2000 and 2005 were around 11 per cent in Nigeria and Vietnam, 2.5 per cent in Indonesia and 1 per cent in Brazil. The Amazon rainforest alone stores about ten times as much carbon as is currently emitted globally per year. Its gradual reduction is therefore leading to substantial additional concentrations of CO_2 in the atmosphere.

Understanding of the climate system and temperature changes is further complicated by various feedback mechanisms. These amplify the initial temperature rise caused by rising levels of greenhouse gases in the atmosphere. These feedbacks can occur quickly or slowly. One example of a fast feedback is that snow and ice melt as a result of rising concentrations of greenhouse gases warming the planet. Le Treut and his colleagues explain that this melting, in turn, reveals land and water previously covered by snow and ice. These darker surfaces 'absorb more of the Sun's heat, causing more warming, which causes more melting, and so on, in a self-reinforcing cycle' (Le Treut et al., 2007, p. 97). Examples of slower feedbacks include the movement of forests and shrubs towards the poles and into tundra regions because of global warming. The expanding vegetation absorbs more sunlight than the tundra areas had done and so adds to the warming of the environment. Another slow feedback is the increasing wetness of the Greenland and West Antarctica ice sheets during the warm season. As tundra melts, powerful greenhouse gases such as methane 'bubble out'. Paleoclimatic records suggest that 'the long-lived greenhouse gases methane, CO_2, and nitrous oxide – all increase with the warming of the oceans and land. These positive amplify climate change over decades, centuries, and longer.' (Hansen, 2008, p. 8).

Alongside and accompanying the feedback mechanisms is the inertia of oceans and ice sheets: a second fundamental element of the earth's climate system. There is a considerable time gap between emissions and temperature change. While it does not take much time for people to emit greenhouse gases and for these to become concentrated in the atmosphere, these gases remain there for periods between several centuries and several thousand years. Some of the effects of temperature increases, such as rising sea levels, can take centuries to appear. Ice sheets also change slowly but can disintegrate within decades. The inertia of the climate system means that only a fraction of the expected surface warming corresponding to the amount of greenhouse gases stored in the atmosphere since the Industrial Revolution has occurred to date. Consequently, the UNDP (2007, p. 4) regards the recognition of the 'combined force of inertia and cumulative outcomes of climate changes' as the 'starting point for avoiding dangerous climate change'. The huge time lags between the causes and the consequences of greenhouse gas emissions are one of the greatest obstacles to effective climate change mitigation strategies. Though the world's climate scientists agree that immediate action is necessary, the climate system's inertia means that the consequences of previous and current greenhouse gas emissions are not yet evident for everyone to see. Yet, if action is delayed (on the grounds that the consequences of climate change do not seem to be too worrying for the time being), greenhouse gas concentrations will continue to build up, and 'by the time the consequences are apparent, the conditions for further temperature increases will already have been created' (Stern, 2009, pp. 16–17). Hence, just as our generation lives with the consequences of greenhouse gas emissions produced since the nineteenth century, 'people living at the start of the 22nd century will live with the consequences of our emissions' (UNDP, 2007, p. 4).

Climate scientists such as James Hansen are unequivocal about the fact that the consequence of the 'combination of inertia and feedbacks is that additional climate change is already "in the pipeline": even if we stop increasing greenhouse gases today, more warming will occur' (Hansen, 2008, p. 8). Hansen further indicates that this combination has brought the world close to 'tipping points', where the climate system may go out of control, 'because the climate state includes large, ready positive feedbacks provided by the Arctic sea ice, the West Antarctic ice sheet, and much of Greenland's ice' (Hansen, 2008, p. 9). 'Little additional forcing' is needed in order to transition to an environment 'far outside the range that has been experienced by humanity, and there will be no return within any foreseeable future generation' (Hansen, 2008, p. 9). Due to the cumulative nature of the climate system, mitigation efforts today 'will not produce significant effects until after 2030', as changing emission pathways do not produce an immediate response in climate systems. As a corollary, the oceans, which have absorbed about 80 per cent of the increase in global warming, 'would continue to rise, and ice sheets would continue melting under any medium-term scenario' (UNDP, 2007, p. 36). Climate scientists are in agreement that any 'wait and see and clean up the mess post factum' approach will not work 'because of inertial effects, warming already in the pipeline, and tipping points. On the contrary, ignoring emissions would lock in catastrophic climate change' (Hansen, 2008, p. 12). Greenhouse gas emissions must, therefore, fall considerably below current levels, whereby the 'rate of emissions reduction required to meet any stabilization goal is very sensitive to the timing and the level of the peak in global emissions' (UNDP, 2007, p. 36).

Future temperature increases will depend on the point in time at which stocks of greenhouse gases stabilise. Whatever the ppm level, this stabilisation requires that 'emissions must be reduced to the point at which they are equivalent to the rate at which CO_2 can be absorbed through natural processes, without damaging the ecological systems of the carbon sinks' (UNDP, 2007, p. 34). The longer emissions remain above this level, the higher the stabilisation level of accumulated stocks of greenhouse gases in the atmosphere. In many ways, climate science relies on projections and simulations, and any predictions about future climate are subject to uncertainties and errors with regard to climate modelling. In order to understand these effects better, the scientific community organised a 'series of systematic comparisons of the different existing models' (Le Treut et al., 2007, p. 117), taking into account factors such as economic growth, population increase and energy-use patterns, including the possible expansion of low-carbon technologies and the development of regional inequalities. Expressed in 'low' and 'high' scenarios, the IPCC (2007a) expects surface temperatures to rise by between 1.1 and 6.4°C by the end of the twenty-first century, with land areas and high northern latitudes warming more rapidly than the global average. Sea levels are expected to rise between 18 cm and

59 cm.[4] The IPCC and the European Commission have both stated that global warming should be limited to 2°C compared to the pre-industrial period. Beyond this threshold, a move towards irreversible damage of the environment and of human development would be difficult to avoid. To have a 50–50 chance of limiting temperature increase to this level would require stabilisation of greenhouse gases concentrations in the atmosphere at around 450 ppm CO_2 equivalents (CO_2e) (Giddens, 2009, p. 22).

In contrast, stabilisation at 550 ppm CO_2e would raise to 80 per cent the probability of going beyond the 2°C threshold, while stabilisation levels at more than 750 ppm CO_2e are associated with temperature changes in excess of 5°C (UNDP, 2007, p. 7). With regard to the possibility that nothing substantial is done about climate change and that decision-makers follow 'business as usual' approaches, Stern (2009, p. 26) estimates that atmospheric CO_2 concentration levels will reach between 750 ppm and 850 ppm CO_2e by the end of the century. In the event of stabilisation of CO_2e at 750 ppm, he estimates a 50 per cent chance of temperature increases of more than 5°C relative to 1850 during the first half of the next century. For 850 ppm, which Stern (2009, p. 26) regards as 'perhaps more likely' if 'business as usual' predominates, 'the probability of being above 5°C would be around 70%, with around a 40% chance of being above 6°C'. Since the publication of the IPCC's fourth assessment report in 2007, and given the cumulative nature of past and current emissions, climate experts have questioned the theory that atmospheric CO_2 concentrations can be held below 450 ppm (Stern, 2009, p. 27). Equally distressing is the suggestion of Hansen and his colleagues that 'the dangerous level of CO_2, at which we will set in motion unstoppable changes [...] may be less' than 450 ppm (Hansen et al., 2008, pp. 11–12). Slow climate feedback processes such as ice sheet disintegration, vegetation migration and greenhouse gas release from soils, tundra or ocean sediments are 'not included in most climate models', but nevertheless 'may begin to come into play on time scales as short as centuries or less'. In consideration of slow climate feedbacks, Hansen et al. (2008, p. 117) conclude that the 'long-term CO_2 limit is in the range 300–500 ppm'. Given the fact that CO_2 atmospheric concentrations are already approaching 400 ppm and given the long lifetime of CO_2, these authors regard a quick transition towards a society approaching zero greenhouse gas emissions to be an essential requirement for avoiding uncontrollable climate change.

Temperature scenarios do not portray potential human development impacts. Some of the effects of climate change are already being felt today. IPCC publications assert that the following phenomena are occurring: enlarged glacial lakes, accelerated rates of melting in permafrost areas in Western Siberia and the Arctic region, acidification of the oceans, retreat of rainforest systems, run-offs from glaciers and snow-fed rivers taking place earlier in the year, earlier onset of spring and the movement of some plant and animals species towards the poles. Current observations of Greenland and Antarctica show 'increasing surface melt, loss of buttressing ice shelves, accelerating ice streams, and increasing overall mass loss' (Hansen et al., 2008, p. 221). The consequences are very different for developed and developing countries. For the developed world, coping with climate change to date has largely been a 'matter of adjusting thermostats, dealing with longer, hotter summers, and observing seasonal shifts' (UNDP, 2007, p. 3). Though there is already an increased risk of flooding in cities like London or Rotterdam as sea levels rise, their inhabitants have so far been protected by sophisticated flood defence systems. Some areas, like Scandinavia, may even benefit in the short term, thanks to less severe winters and improved opportunities for agriculture or sea routes. However, the heat wave of 2003 in Europe, when the number of people who died was 35,000 over what is usually the case, recurrent water shortages in Southern Europe and the disaster of Hurricane Katrina in New Orleans in 2005 appear to be harbingers of what the future may hold for the Atlantic space (UNDP, 2007, pp. 81–4). Of course, no single natural disaster can be attributed to climate change, but climate scientists agree that climate change is generally increasing the risk and the frequency of such tragic events. The impact of climate change on the developing world has so far been much more severe than on rich, industrialised countries (UNDP, 2007, pp. 85–115). It has already affected weather patterns in Asia and Africa, where droughts occur more frequently. As glaciers disappear, several regions on these continents are losing their primary dry season freshwater sources, so that, on average, more time must be spent and greater distances must be overcome in order to collect water. Crops also fail more often, so that undernourishment occurs more regularly. Rural poverty increases, as does the infection rate of diseases such as malaria. Annually, 120 million people are exposed to tropical cyclones (IPCC, 2007a). The Global Humanitarian Forum (2009, p. 1), the think-tank of the former UN Secretary-General Kofi Annan, put at 300,000 the number of people whom climate change is already killing annually, and the number of those seriously affected at 325 million. The vast majority of the affected live in developing countries.

It goes without saying that predictions about the potential long-term impacts of temperature increase are less reliable, since these depend on the magnitude of the increases and on the nature of feedbacks. These feedbacks, in turn, may trigger 'tipping points' – such as an ice-free Arctic – that are difficult to include in climate modelling. However, as the temperature rise is very likely to be 2°C at least, the general expectation is one of greater variability in climate conditions (Eriksen et al., 2007, p. 3). Hot summers will be followed by cold spells, heavy monsoon rainfalls by intensely dry periods. Increased extremes of temperatures are likely to be accompanied by more frequent precipitation events, with more areas affected by drought. The melting of glaciers and snow is likely to lead to torrents and flooding during wet seasons (Stern, 2009, p. 29). Climate scientists expect a great expansion of dry areas in Africa and Southern Europe. Likely consequences of a temperature rise of 2°C further include an increase in the number and intensity of tropical cyclones, storms and hurricanes and in the incidence of extremely high sea levels and floods. While the IPCC (2007a) anticipates that sea levels would rise in the range of 18–59 cm over the next century, more recent research suggests that, since sea levels react very slowly to temperature increases, the 'eventual rise might be very much larger and estimates go as high as several metres per degree centigrade of warming' (Stern, 2009, p. 28). Asian countries will be most affected, while flood defences in Europe are seen as comparatively solid. If the temperature rises by 2–3°C over the next decades, Stern (2009, p. 29) asserts that there is a high risk of severe dislocation as a result of rising sea levels and of the collapse of major rainforests.

It is even more difficult to imagine the far from unlikely case of a world with temperatures 4–5°C above current levels. To do so, we would need to go back about 30–50 million years – to the Eocene period – when this was the temperature of the earth (Stern, 2009, p. 31). At any rate, the likelihood of catastrophic impacts would be much higher due to the heightened risk of feedbacks that could trigger tipping points in the climate system. While some areas, for example in Southern Europe, could very likely turn into deserts, most of Florida and Bangladesh would 'eventually be submerged' (Stern, 2009, p. 31). The last time when temperatures were 4 to 5°C lower was during the last ice age, around 10,000 years ago, when much of Europe and North America was covered by more than one kilometre of ice and 'human beings were concentrated much closer to the equator than now' (Stern, 2009, p. 31). What is completely unprecedented and unique with regard to the prospective temperature change is its much faster pace by comparison to the great stretches of evolutionary time during which previous temperature changes occurred. A rise in temperature of 4–5°C or more over a few decades is well beyond human experience and imagination. One thing, however, is clear: temperature changes of this magnitude over such a short period of time mean that the 'physical geography is rewritten'. And, if this occurs, 'so too is the human geography of the world' (Stern, 2009, p. 31). Billions of people would migrate at short notice, plunging 'the world into massive and extended conflict' (Stern, 2009, p. 31). Other scholars plausibly expect 'resource-based wars' to 'dominate the current century' (Giddens, 2009, p. 22), due to mass destitution and migration caused by flooded coastal cities and/or desertified rural areas. Again, given their location and lack of resources, the developing countries would be more seriously affected than the developed countries. The long-term development opportunities of developing countries would be undermined, placing 'further stress on already over-stretched coping mechanisms and trapping people in downward spirals of deprivation' (UNDP, 2007, p. 8).

The temporary easy ride that most developed countries have enjoyed with regard to the impacts of climate change should not tempt their elites to be idle and to follow business as usual. Climate scholars are in agreement that 'all countries', including the rich, 'will have major challenges in adapting to rapid climate change. Those that experience some offsetting benefits in the short term will also have major short- to medium-term adjustment cost' (Stern, 2009, p. 30). Yet it is questionable whether the elites of the rich countries will indeed act responsibly, given the enormous challenge that climate change presents, since the climate cycle does not coincide with political cycles and election periods. The long-term interests of the planet are not the same as the short-term interests of our 'leaders'. There is at least one parallel between the climate system and society, in that both are characterised by 'inertia', due to which changes proceed slowly if at all. Policymakers, like many other people, are often not aware of the nature and the social genesis of the economic, political and ideological structures within which they operate – let alone the consequences of their decisions – and they tend towards policy responses that are in harmony with established patterns. In policy circles, this is known as 'path dependency'. The room for developing alternative policies is normally small, and any policies implemented are likely to intensify the institutional network they came from – thus amplifying the anomalies that are outcomes of the network. Many research projects into socio-economic regulation indeed demonstrate that the nature of particular political pathways is bound up with power relations and specific group interests, which are expressed, among other things, in the dominant patterns of economic reasoning, political regulation and social inequality, both at the national and at the international level. I would hypothesise that climate policies do not constitute an exception to this rule.

It is not only politicians, but also social actors who generally tend to be unaware of the powerful social structures within which they operate, but which have great impact on daily life. Insight into these structures facilitates understanding as to why change in modern societies and in individuals' behaviour proceeds slowly, if at all. Thus the process of revealing social structures not previously discerned also helps to identify the conditions required in order to bring about such change and to overcome these structures.

One of the major social structures of contemporary society, which developed almost simultaneously with the issue of climate change, is capitalism. Capitalism is not just a particular way of organising the division of labour. This organisation corresponds to particular social relations (including power relations), discourses and ideologies with which people make sense of society. As political economists of various persuasions have shown, capitalism develops in distinct stages or growth strategies and functions satisfactorily – in economic terms – only when particular ways and methods of producing commodities are in harmony with suitable modes of consumption. Since these stages take on different forms over time, the present book addresses the homology between particular periods of capitalist development and climate change. It sets out to delineate both continuities and ruptures in the ways in which particular capitalist growth strategies have affected the climate system. The book's structure results from this particular research interest.

The book first identifies and discusses the general tensions between the imperatives of capitalist production and reproduction and those of the earth as an ecological system – a tension that is regulated in different ways during different capitalist periods of growth (Part I). Subsequently it addresses the parallel development of the two most recent growth strategies – Fordism and finance-driven capitalism – and the issue of climate change (Parts II and III). Part IV analyses the framework of international climate regulation against the background of huge and persistent global inequalities, with special emphasis on the theory and practice of free-market approaches towards climate change policies.

Part I
Capitalist Development and the Regulation of Society and Nature

This first part begins with a general reflection on the interaction between nature and its inherent reproductive principles and human societies through the labour processes (Chapter 1). Chapter 2 analyses the particular forms this interaction takes in capitalism. At a fairly high level of abstraction, by building on the work of Karl Marx and ecological economists, it discusses the impacts on the environment, and on the climate in particular, of a mode of production in which labour products take the form of commodities and the production process is geared towards maximising exchange value. Chapter 3 introduces the regulation approach as a means of understanding the changing institutional arrangements of different capitalist growth periods, their consumption models and energy regimes. The regulation approach employs 'intermediary concepts' such as 'accumulation regime' and 'mode of regulation', which will be applied to develop a social science contribution towards understanding climate change.

1
Nature and the Work Process

Much of modern economic theory proposes a circular flow of exchange. Economics is seen as a repetitive cycle linking money and commodities with households and companies. It is understood as being circular and reversible: a 'return to capital' basically means that the original capital spent, augmented by a surplus, returns to its owner and the process of capital valorisation starts all over again, on a greater scale. Yet the circular monetary value aspect of economics is coupled with a physical flow and throughput of matter and energy, which is ultimately linear. And, although it is 'the linear throughput, not the circular flow of value, that impinges on the environment in the forms of depletion and pollution', it is the circular flow that 'has the spotlight [in economic theory], while the concept of throughput is only dimly visible in the shadows' (Daly, 1985, p. 280). In neoclassical theory especially, the production of goods and services is analysed from the standpoint of the growth of monetary value, which is seen as indefinite, while the roles played by energy and natural resources in this production are usually not mentioned. Hence this type of economic analysis tends to finish at the point where the flows of money stop: 'the goods and the services produced by human activity only appear in the economic system insofar as they exist in the form of commodities, and they drop out of sight as soon as they lose this quality' (Deléage, 1994, p. 38). If the issue of resources is discussed at all, it is assumed that 'substitution' processes will sort out the problem of depletion. Indeed, in the case of 'perfect competition', which is an essential requirement in neoclassical growth models, 'the price system will see that "correct" substitutions are made at the right times' (Miernyk, 1999, p. 75).[1] While the neoclassical perspective represents the economy as a closed system – within which flows of services and goods are compensated by financial flows in the opposite direction and

17

coherence is guaranteed through the link of exchange alone – matter, energy and nature, in general, have largely been treated as if they were infinite.[2]

Economics, however, has not always been seen as synonymous with a science of prices and economic value (De Gleria, 1999, p. 84). In the physiocratic system, for example, the notion of natural resources was central. The wealth of nations was derived solely from the value of land, and the entire economic process was understood through focusing on a single physical factor: the productivity of agriculture was the only kind of work that created value and surplus. The Physiocrats developed an original theory of surplus by arguing that agricultural work alone yielded a surplus over cost. Hence, in the Physiocratic model, economic profit was derived from 'unrecompensed work done by nature, since in setting food prices, cultivators take into account their labor and expenses as well as the surplus value contributed by the fertility of the soil' (Cleveland, 1999, p. 127). The classical tradition of Adam Smith and David Ricardo moved the focus away from agricultural production, pointing out that it is labour that produces exchange value. Karl Marx, for his part, also aimed to show that surplus originates only in labour. But these authors did not go as far as removing nature and its material resources from the analysis of economic processes. On the contrary, they all agreed with the seventeenth-century English economist and philosopher William Petty, who thought labour was not the only source of material wealth. Petty saw labour as the 'father' of material wealth and the 'earth its mother' (cited in Marx, 1961, p. 43). Far from ignoring the role of nature in the productive process, Marx (1961) expressed the classical view that 'useful labour [...] is a necessary condition, independent of all forms of society, for the existence of the human race; it is an eternal nature-imposed necessity, without which there can be no material exchange between man and Nature, and therefore no life' (p. 43). This abstraction of both value and surplus from labour and from the role of nature in the process of wealth production was only attained when the subjective determination of value became privileged in neoclassical economics. Surplus disappeared from the analysis due to the attribution of the whole revenue to productive factors, according to their marginal productivity. Natural resources, for their part, were sidelined or totally ignored – through the introduction of new categories such as decreasing marginal utility and productivity, and by denying 'that they are a different factor from capital' at all (De Gleria, 1999, p. 86).

Thermodynamic economics

The sidelining or removal of natural processes from economic reasoning is a problematic trend, as is indicated by research in both bioeconomics and thermodynamics. Rather than exclusively focusing on the movement of values, the bioeconomic perspective deals with the extraction of raw materials and their elimination in the form of waste – which are considered the 'first and last phases of all economic activity' (Deléage, 1994, p. 40). This view allows the link between nature and economic processes to be addressed at two points: first, when natural raw materials are extracted; second, when they are deposited into the environment in the form of waste. It follows that, if the overall scale of production increases, deposit sites will also grow. And, as more accessible and concentrated deposit sites become exhausted, 'either larger amounts of energy have to be mobilized to tap the less profitable deposits, or new technologies must be developed, requiring a growing effort in innovation' (p. 41). Since nature is unable to recycle all the new waste, the social cost of waste management increases and waste management itself becomes a socially contested issue.

The fact that economic activities cannot ignore the laws of physics is one of the essential insights of thermodynamic economics. Building on the work of Nicholas Georgescu-Roegen (1971) in particular, this perspective stresses that, in production, transport, communication and consumption, processes of irreversible material and energy transformation take place that are normally not considered by economists. This use of the first law of thermodynamics builds upon Einstein's theories on mass and energy and claims that there is a 'conversation' going on between the energy and the material reserves of a system (ultimately, of the universe). The second law captures the fundamental asymmetry of the universe, in which the distribution of energy changes in an irreversible manner. The 'measurement' of total disorder or chaos in a system is 'entropy', and all economic activity runs against the general tendency of the universe to move towards a state of greater disorder, or higher entropy. The overall increase in entropy resulting from production processes is always greater than the local decrease arising from the production of any concrete good. In an isolated thermodynamic system, all energy transformations convert energy into less available and less useful or ordered forms, so that the system is increasingly less able to produce work (in the physical sense).[3] In other words, the portion of free and unbound energy declines by comparison with that of bound and dissipated energy, which can no longer be converted into work. The relevance of the two laws for economics is exemplified by heat, which can only dissipate and will not flow from a cold to a hot object or area.[4] Energy cannot be transformed into work 'without some of the energy being dissipated as unrecoverable heat. An engine cannot operate at one hundred per cent efficiency, that is, on a cycle whose only effect is to convert energy into work: a refrigerator will not operate unless it is plugged in' (Burkett, 2005, pp. 117–18).

The continuation of the work process (in the socio-economic sense), whatever its historical form, is therefore dependent on a continuous input of low-entropy energy for the rearrangement of matter. During the work process, free energy is 'degraded into bound or unavailable (high-entropy) energy', whereby low-entropy raw materials are 'easier to rearrange (use up less free energy) than are dissipated waste materials (slag, sludge, soot, ashes, tailings, rust, etc.)' (Daly, 1974, p. 284). Thermodynamic economists also emphasise that environmental sources and sinks of energy and raw materials are finite, that is, they can be used only once. Hence, the 'entropic nature' of material and energy throughput greatly increases the problem of scarcity, as finite sources tend to 'run down and sinks fill up, and the latter cannot replenish the former' (Daly, 1985, pp. 285–6). Indeed, 'if the entropic process were not irrevocable, i.e., if the energy of a piece of coal or of uranium could be used over and over again ad infinitum, scarcity would hardly exist in man's life' (Georgescu-Roegen, 1971, p. 6). From the thermodynamic perspective, the critique of mainstream economics is precisely that the latter, despite selling itself as the 'science of the "rational use of scarce resources" ', is 'unaware of the thermodynamic basis of its own central category: scarcity' (Altvater, 1994, pp. 83–4).

Entropy as such is inevitable, since matter continues to dissipate and disperse into less ordered and less useful forms. This applies not only to natural organic decomposition and to the corrosion processes of natural forces, but also to the 'various kind of *friction* produced by the material mechanisms needed to convert energy into work' (Burkett, 2005, p. 120). Georgescu-Roegen (1981, pp. 59–60) concludes from this that 'complete recycling is impossible'. Daly agrees with the argument and consequently does not assume that entropy remains constant in a 'steady-state economy'. Instead, he emphasises the minimisation of the matter–energy throughput and the accompanying increases in entropy that such an economy would enable. Hence the advantage of Daly's economic and ecological model over neoclassical growth models is not that it is characterised by 'constant throughput', but that it provides a 'strategy for good stewardship, for maintaining our spaceship and permitting it to die of old age rather than the cancer of growthmania' (Daly, 1974, p. 16). Georgescu-Roegen, too, has criticised the predominant modes of socio-economic development on the grounds that these were primarily geared towards economic growth (in monetary terms); and he reminds us that, due to entropic process, 'every Cadillac produced at any time means fewer lives in the future'. What mainstream economics celebrates as the historical process of 'technological progress' has always been accompanied by the 'shift from the more abundant source of low entropy – the solar radiation – to the less abundant one – the earth's mineral resources' – a transition that, *ceteris paribus*, brings the 'career of the human species nearer to its end' through a 'speedier decumulation of its dowry' (Georgescu-Roegen, 1971, p. 304).

Although the entropic process as such is inevitable, the rate of entropy production varies over time and with regard to the different historical forms of organising the economy. Georgescu-Roegen has identified the different stages in human development with the associated dominant technological paradigms, which he calls 'Promethean' technologies. Through the qualitative conversion of energy, these produce an irreversible change in the relationship between economic agents and nature and cause profound modifications in natural ecosystems and human societies. The three Promethean transitions are as follows: animal husbandry, fire and heat engines, each marking an evolutionary step within human economic development. Georgescu-Roegen further observes that any Promethean technology requires a 'steady diet of natural resources: fire requires wood, steam engines require coal'. Each also serves as 'the basis of an ongoing social system' (Beard and Lozada, 1999, p. 135). While the transition towards agriculture 'moved human societies from living off the direct flows of food from nature obtained by hunting and gathering to living off stocks of stored-up organic material and using the power of other animals', the shift to steam-driven energy precipitated 'more rapid industrialization, providing countless material comforts while leaving much of the world utterly dependent on stocks of fossil fuels' (Mesner and Gowdy, 1999, p. 58). Mesner and Gowdy build on this by arguing that each of these Promethean technologies caused its own environmental crisis due to it being in general use:

The mastery of fire resulted in widespread changes in landscape as pre-agricultural peoples used fire to encourage plant growth and attract game animals. Agricultural practices, which often went in tandem with clearing of forests, also led to loss of valuable topsoil, increasing flows of nutrients into rivers and streams, and water short-ages due to irrigation. The air pollution associated with the steam engine gave rise to a host of new problems – acid rain and global warming to name only two.

Climate change as a socio-ecological issue cannot be understood as a direct outcome of entropy, since, as Schwartzman (2008, pp. 51–3) argues, the latter alone is 'too abstract and coarse a concept'. Yet if one combines, like Georgescu-Roegen, the general concept of entropy with the socio-historical notion of Promethean technologies, which corre-spond to different entropic rates, then a different story can be told. Anthropogenic climate change did not emerge during the era of hunt-ing and gathering or in agricultural societies; it began with the Industrial Revolution, which itself was accompanied by an unprecedented increase in the depletion and burning of fossil fuels, thereby accelerating the entropy rate and the greenhouse effect. Neither could the Industrial Revolution have proceeded without concomitant political and socio-economic upheavals. Hence Georgescu-Roegen made an important step beyond an exclusively natural scientific view of entropy, environmental degradation and climate change, through the concept of 'Promethean technology'. By embedding these technologies into historically diverse socio-economic circumstances, he criticised the most recent economic stage for its reckless consumption of fossil fuels. His famous dictum that 'matter matters, too' serves as a powerful reminder that the deple-tion of resources is not only a physical but also a political and ethical issue. Indeed, following Georgescu-Roegen, current economic processes should be understood and assessed under the categorical imperative to 'preserve resources for the future inhabitants of our world' (Beard and Lozada, 1999, p. 136). The fact that there is 'no substitute for natural resources' is followed by the moral obligation upon our generation to limit 'our use of our terrestrial dowry' (Beard and Lozada, 1999, p. 136).

Knowledge of advanced climate change and other ecological crises underlines the significance of Georgescu-Roegen's appeal for a fun-damental review of our work and consumption processes from an ecological perspective. Yet his analysis, though providing a historically comparative perspective, remains largely at the level of technological – that is, Promethean – innovations. His critique of contemporary pro-duction and consumption norms remains situated at the normative or moral level and fails to address their embeddedness into specific economic categories, social relations and the associated discourses and ideologies. In consequence, Georgescu-Roegen and his followers lack the analytic tools to provide theories on the specific socio-economic relations from which the current lack of moderation in fossil fuel consumption results (which, in turn, is causing the depletion of fossil fuel reserves). I would, therefore, agree with Paul Burkett's criticism that ecological economics has recognised the anthropomorphic character of entropy and associated environmental crises, but has failed to provide the tools 'with which to critically analyse this contingency in particular economies and societies' (2005, p. 130). Consequently, Burkett insists on linking the 'instrumental and/or functional goals' of production and consumption with the 'social relations of production' (p. 131). Here I concur with Burkett – and also with other recent eco-Marxist authors: in contrast to most ecological economists, who fail to take this analytical step, the Marxist tradition provides the tools for making the analysis and the link advocated by Burkett.

As far as the general work process is concerned, Georgescu-Roegen and more recent ecological economists do not deviate from Marx's time-honoured principle that 'all epochs of production have certain common traits, common characteristics' (1973, p. 320). No production is possible 'without an instrument of production, even if this instrument is only the hand'. Throughout the centuries, even if we go back to 'simple exchange or barter', the purpose of production has always been the production of use values (pp. 267–8). Closely associated with production at this most abstract level is the general concept of labour and of the work process, which is itself directly linked to nature: 'a process in which both man and Nature participate, and in which man of his own accord starts, regulates, and controls the material re-actions between himself and Nature' (Marx, 1961, p. 177). For Marx, therefore, labour is the connecting link between nature and human beings, who, in order to survive, must interact with it and transfer natural raw materials into use values. By analogy with the work of Justus Liebig and other German physiologists during the 1830s and 1840s, who had employed the concept of 'metabolism' in relation to tissue degradation (Burkett and Foster, 2006), Marx characterised the interaction between human beings and nature in the labour process as the 'necessary condition for effecting exchange of matter between man and Nature; it is the everlasting Nature-imposed condition of human existence' (Marx, 1961, pp. 183–4). While Marx and Georgescu-Roegen could easily have agreed on the fact that this 'metabolic' relation constitutes the universal condition upon which human life is sustained, the latter would part company when the former asserted that the notion of 'production in general' is no more than 'a rational abstraction in so far as it really brings out and fixes the common element and thus saves us repetition' (Marx, 1973, p. 320). For Marx, real production processes and the associated relationship between the economic agents and nature take place in specific social forms; and it is the particular features that this relationship takes in the capitalist mode of production, during which climate change emerged and became more serious, that we will now turn to.

2
Capitalism, Nature and Climate Change: A Structural Analysis

In a letter to Ludwig Kugelmann (11 July, 1868), Marx clarifies his viewpoint that the work process is both the point and the form of the interaction (or the 'metabolic' relation) between humans and nature. First, he maintains that the performance of labour is necessary for and central to human existence and, second, he makes clear that no social form of production can escape natural laws. In relation to the first proposal, he suggests that 'every child knows that any nation that stopped working, not for a year, but let us say, just for a few weeks, would perish' (Marx, 1988, p. 67). And, as 'every child knows, too', social labour needs to be distributed 'in specific proportions' in order to match the 'differing amounts of needs'. Labour is the form in which humans interact with nature through the exchange and transformation of organic matter. As economic agents, humans 'both confront the nature-imposed conditions of the processes found in the material world and affect these processes through labour' (Clark and York, 2005, p. 398). The ways in which social production are organised change, but these social transformations 'can only change' the 'form of manifestation' of natural laws in the work processes. 'Natural laws cannot be abolished at all'. Hence, the 'only thing that can change [...] is the form in which those laws assert themselves'. And the specific form in which the division of labour asserts itself as 'private exchange of the individual products of labour, is precisely the exchange value of these products' (Marx, 1988, p. 67, originally 1868).

Hence the specific nature of the capitalist mode of production does not lie in the fact that labour is performed and use values are produced in collective ways, but in the fact that these processes are mediated through commodities and money. Goods are normally not produced for one's own consumption but for the purpose of selling them on anonymous markets. Before a use value can be consumed, it must prove itself as an exchange value. Whenever labour products take the form of commodities, the division of labour and its distribution 'in specific proportions' proceed in indirect ways, that is, mediated through exchange operations. And, since economic agents do not come in contact with each other directly, or prior to the exchange of goods (before or during their production), but only indirectly and only through the exchange of their commodities, Marx does not begin *Capital* with the analysis of wage labour, exploitation or accumulation, but with an examination of the commodity and its twofold character, as use value and exchange value. For Marx (1961, p. 41), this was indeed the 'pivot on which a clear comprehension of Political Economy turns', since the basic contradiction between use value and exchange value expresses at the most abstract level all the contradictions and tensions to be found in further economic categories, social relations and modes of consciousness at lower levels of abstraction.[1]

Use value, exchange value and the carbon cycle

Exchange value refers to the commodity's monetary value for the seller, while use value is concerned with the commodity's material and/or symbolical usefulness for the purchaser. Both are interdependent: 'Without exchange value, commodities would not be produced for sale; without use value, they would not be purchased' (Jessop, 2002, p. 16). In contrast to neoclassical economics, which tends to ignore the material aspect of the production and distribution of goods, Marx emphasises the twofold character of the commodity, thus building directly upon the difference between value and matter. Yet, although an ecological view and the consideration of environmental issues in economics hinge upon this use-value element, the simultaneous existence of the commodity as exchange value results in the abstraction 'from the fact that materials and energy are transformed through concrete, quality-changing labour' (Altvater, 1993, p. 190). Under the imperative of valorisation, the concrete and material aspect of labour, which is reflected in use values of commodities, is subordinated to abstract labour and exchange value and, hence, somewhat sidelined. Use values, matter and energy are not of primary interest on capitalist markets; of interest is instead their form as values – that is, as repositories of abstract, socially necessary labour. Far from being flawed, Marx's analysis, which dedicates more space to the value than to the use-value aspect of the commodity, accurately reflects capitalism's tendency towards underevaluating use values and, hence, 'natural conditions' in general (Burkett, 1999, p. 19). James O'Connor (1998, p. 327) illustrates the subordination of concrete labour and use value to abstract labour and exchange value by referring to the direct negative impacts of this subordination on the environment. Ecological goals such as sustainable land-use practices, the preservation of species diversity and, in the sphere of consumption, clean air and water and non-congested transport networks are governed 'first and foremost by the need to produce exchange value' – that is, their societal handling has to respect the priority of valorisation. Capitalist 'rational' cost–benefit analyses ignore ecological issues and natural limits 'as long as these do not make themselves felt as economic limits – that is, as a cost burden within the economic system' (Altvater, 1993, p. 184).

At the next step, Marx (1961, pp. 94–142) compares two kinds of exchange, or two 'metamorphoses' between commodity and money. In the first one – commodity–money–commodity – the purpose of the exchange is qualitative. A holder of a commodity exchanges it for its money equivalent, then he or she buys another commodity for his or her own use: 'selling in order to buy' (p. 147). In this exchange, the role of money is that of a measure and store of value as well as that of a legal tender. Now Marx (p. 130) argues that, with the 'very earliest development of the circulation of commodities, there is also developed the necessity, and the passionate desire, to hold fast the product of the first metamorphosis': money serves here as the general and ultimate expression of the wealth available in a society, or as capital that potentially leads to profit and bears interest. The purpose of the second metamorphosis – money–commodity–money – can only be a quantitative one, since there is no qualitative difference between its origin and its result: the production of more money as a materially homogenous entity as compared to the original amount. The natural and material ingredients of the production process, in contrast, are heterogeneous. Their combination in the work process is based on qualitative changes in the form of rearrangements of energy and matter. A further tension between the monetary form of values and the principles of natural reproduction is that the former is completely divisible in terms of monetary quota, while the latter, natural world, of which the work process is composed, represents 'highly interconnected and interdependent material, biological and thermodynamic systems of varying entropy levels'. Furthermore, Burkett argues that monetary claims on wealth in the form of bank accounts, stocks or bonds are highly mobile, and this often contradicts the locational fixedness and specificities of ecosystems. Finally, while money and valorisation are quantitatively unlimited and hence reversible, low-entropy matter and energy are not. The earth's stock of fossil fuels, in particular, is confined, and the existing stock can only be burnt once. In other words, it is irreversible. Burkett concludes that money and its accumulation are 'homogenous, divisible, mobile, reversible and quantitatively unlimited, by contrast with the qualitative variety, indivisibility, locational uniqueness, irreversibility and quantitative limits of low-entropy matter-energy'. There is a structural tension between the value and money form of societal wealth and its material and energy substance (Burkett, 2005, p. 144).

Marx's theory of how a profit is realised on the grounds of the principle of exchanging labour equivalents is well-known. Profit production is possible due to the fact that a commodity is available for sale that has the use value of creating exchange value and that can be used longer than that which represents the cost of its own reproduction: labour power. Capitalism is characterised as being a mode of production where producers – as wage-earners – are largely separated from their means of subsistence and production and have no alternative but to offer the only commodity at their disposal on 'labour markets'. The other 'factors of production' – land, raw materials, fuels, auxiliaries and so on – can be purchased on separate markets, and it is only through the intermediation of employers, who hold the necessary capital, that the former comes in contact with the latter. Hence capitalist production 'can spring into life, only when the owner of the means of production and subsistence meets in the market with the free laborer selling his labour power. And this one historical condition comprises a world history' (Marx, 1961, p. 170).[2] This implies that capitalism's reproduction requirements are distinct from the material and ecological preconditions for the reproduction of labour power and of the other factors of production. For capitalist production, all that matters is that these factors and the ingredients of material production are separately available for purchase, and in forms that can be combined in the production process of capital. Given this precondition, capitalist reproduction tends to disrespect the imperatives of natural reproduction, for instance to preserve the fossil fuel stocks, on account of its inherent tendency to expand the scale of production (see below). James O'Connor (1988) makes an important point when he theorises the tension between the logic of capitalist reproduction and that of its conditions of production in terms of a 'second contradiction of capitalism',[3] even if one may not want to go as far as this. The conditions of production involve not only the particular social, cultural and ecological conditions of the reproduction of labour power, but also those of capitalism's natural environment: forests, oil fields, water supplies, a functioning atmosphere and so on. One thing these 'conditions' have in common is that they were not produced by capitalist methods, therefore Karl Polanyi (1944) referred to them as 'ficticious commodities'. O'Connor argues that the tendential degradation of these conditions does not make itself felt directly, as a factor of crisis, but indirectly, through rising costs that diminish profits and constrain the supply side: 'Limits to growth thus do not appear, in the first instance, as absolute shortages of laborpower, raw materials, clean water and air, and urban space, and the like, but as high-cost laborpower, resources, and infrastructure and space' (O'Connor, 1988, p. 243). Cost crises originate when individual companies follow growth strategies that, in time, 'degrade or fail to maintain over time the material conditions of their own production, for example, by neglecting work conditions (hence raising the health bill), degrading soils (hence lowering the productivity of land) or turning their backs on decaying urban infrastructure (hence increasing congestion costs)' (O'Connor, 1994, p. 162).

At the level of the capitalist production process, the tension between monetary values and material or natural use values is reproduced in greater dimensions. Marx (1961, pp. 312–21) discusses the link between the valorisation and the expansion of the scale of production when analysing the production of 'relative surplus'. The profitability of a company can be improved not only by increasing the working hours of the wage-earners ('absolute surplus value'), but also by shortening the part of their working day that is necessary for the workers' physical and social reproduction. Marx explains a reduction in the price of labour power (and, all other things being equal, the magnification of the employers' profit) through increases in productivity in those branches of production that are part of the consumption patterns of the wage-earners. However, he also stresses, following his assumption that labour is the only source of value, that the realisation of such a relative surplus value will face an imminent contradiction: individual owners of capital are permanently motivated to optimise the technological and organisational basis of the work process in order to be one step ahead of their competitors. This is normally carried out through a substitution of workers by machinery or through an improved organisation of the internal division of labour. The employers whose productivity level is above average can thus make extra profit, since they are able to sell their commodities at prices below the normal level. Yet such an improvement of production methods tends towards generalisation, and the extra profit moves towards zero, since competing companies have no choice but to copy the new methods, or even to improve upon them. Since this improved level of productivity gradually becomes the new social standard, a given quantity of commodities is now produced with less labour effort than previously. The price of a single commodity decreases as a result. Marx concludes that the methods of production of relative surplus result in workers being driven out of the production process. On the one hand, the rate of surplus of the employed workers increases; on the other hand, however, the absolute volume or mass of surplus value (and, other conditions being equal, the mass of profit) decreases, since fewer workers are needed to produce a given amount of commodities than before. In order to keep the volume of profit stable despite this dilemma, there is no alternative but to expand the overall scale of production through the reinvestment of previous profit; in other words through 'accumulation'.

Next, in the partially historical chapters on cooperation, the manufactory and 'machinery and modern industry', Marx (1961, Part IV) discusses the advancement of the division of labour and how the work process became independent from the subjective limitations of individual workers through the systematic application of natural forces and the natural sciences. The Industrial Revolution introduced tools and machinery that reduced the role of many individual workers to that of an 'appendage'. When the work process had an industrial foundation, the subjugation of nature under capital became more complete. Now nature was 'for the first time [...] purely an object for humankind, purely matter of utility; ceases to be recognized as a power to itself; and the theoretical discovery of its autonomous laws appears merely as a ruse so as to subjugate it under human needs, whether as an object of consumption or as a means of production' (Marx, 1973, p. 409). Marx also shows that expanding scales of production, which are a corollary of the valorisation logic of capitalist production, normally coincide with greater amounts of throughput of raw materials and auxiliary substances, especially in the form of fossil fuels and of available energy. All other things being equal, an increase in productivity means that a given work force processes a larger quantity of raw materials and consumes more energy. Rising demand for raw materials and available energy normally leads to rising prices, for example for crude oil, creating incentives for individual companies to recycle and to use a given quantity of materials or fuels in more efficient ways. Marx (2006, Chapter 5) described this as a long-term trend towards a greater 'economy in the employment of constant capital'. Yet progress in the efficiency of raw and auxiliary materials does not fundamentally alter the link between the expansion of the scale of production and the increase in the material and energy throughput, a phenomenon that had been observed by the British economist William Stanley Jevons in the 1860s (Jevons, 1865). According to the 'Jevons paradox', greater efficiency in the use of a fossil energy source such as coal or oil leads to an increase in demand – not to a decrease. On the contrary, this increase becomes the precondition for further capital expansion (see Foster, 2009, p. 124).

Brett Clark and Richard York (2005, pp. 402–3) have discussed in the most systematic way the calamitous ways in which capital accumulation affects the carbon cycle, on which all life forms ultimately depend (see Introduction). While over the 'past 400,000 years, the carbon cycle and climate system have operated in a relatively constrained manner, sustaining the temperature of the earth and maintaining the balance of gases in the atmosphere', the accumulation of CO_2 in the atmosphere of the most recent 160 years has led to a 'rupture of the carbon cycle'. It was indeed

not until the rise of capitalism, and especially the development of industrial capital, that anthropogenic CO_2 emissions greatly expanded in scale, through the burning of coal and petroleum, exploiting the historic stock of energy that was stored deep in the earth and releasing it back into the atmosphere. As a result, the concentration of CO_2 in the atmosphere has increased dramatically, overwhelming the ability of natural sinks – which have also been disrupted by anthropogenic forces – to absorb the additional carbon and leading to climate change.

(Clark and York, 2005, p. 403)

Due to the contradiction inherent in relative surplus production, capitalist development is bound up with the expansion of the scale of production, and, in order to achieve this, 'existing barriers, both social and natural (such as operating within the regulative laws of natural cycles)' are constantly being transcended, while at the same time 'new barriers (such as natural limits and rifts in metabolic cycles)' are being created in the continuous search for profit (p. 407). By plundering the 'historical stock of concentrated energy that has been removed from the biosphere only to transform and transfer this stored energy (coal, oil, and natural gas) from the recesses of the earth to the atmosphere in the form of CO_2' (p. 409), capitalism has been disrupting the carbon cycle at an accelerating rate.

Finally, in the second and third volumes of *Capital*, Marx demonstrates that capital does not exist only in its productive, that is, value-producing, form, but also in unproductive forms, as money and commodity capital. While alternating between these three forms, competition forces individual companies to reduce the two unproductive functions of the capital cycle and, hence, to speed up the overall turnover process as much as possible. Temporal differences between input and output of the production process are diminished as a result. Similarly, geographic distances between different production stages and locations of production and sales, which delay both the production and the sale of commodities, are reduced through the improvement of infrastructure (roads, bridges, railways, airports, harbours and so on) and of communications (telegraph, telephone, Internet and the like). Yet, although long-term capitalist development with regard to its value side is characterised by 'time–space compression' (Harvey, 1990), we have already noted that the transformations of material and energy that characterise the use-value side of commodities are linear and irreversible. The tendency towards temporal and geographic 'simultaneity' suggested by the logic of valorisation, and in particular by the logic of increasingly rapid turnover cycles, is indeed 'unattainable' and in contradiction to the irreversible and linear character of all work processes, which, regardless of their social forms, follow a ' "time-arrow" moving from the past through the present to the future' (Altvater, 1993, p. 200) and are accompanied by a decline in the quality of the respective matter and energy. We might conclude with Altvater (1993, pp. 201–2) that the 'quantitivism and expansionism' that characterises the economic system is responsible for the fact that the 'whole planet is subordinated to the capitalist principles of the transformation of values and materials, so that it becomes increasingly inadmissible to postulate open systems with an environment rich in energy and materials and to ignore irreversibilities'.

Table 2.1 summarises the argument concerning the tensions between nature and the reproduction principles of capital at the different levels of abstraction taken in *Capital*. Far from ignoring material and natural processes in the economic process, Marx's emphasis on the double nature of the commodity and of work – as value in motion and as a concrete stock of invested time- and place-specific assets of matter and energy – renders his critique of political economy amenable to ecological laws. The subsequent discussion – on money and capital, labour power and the conditions of production, the advancement of the division of labour through industrial capital as well as the accumulation and turnover process of capital – further illuminates the contradictions

Table 2.1   Tensions between capitalist development and nature

Form	Exchange-value moment	Use-value moment
Commodity and production – Commodity's double character serves as 'pivotal point' for the critique of political economy (and ecology)	– Exchange value reduces concrete works as well as matter and energy to repositories of abstract labour	– Use-value and work-process moments allow for a consideration of ecological parameters in economic analysis, which are sidelined both in the exchange side of the commodity and in neoclassical reasoning
Money and valorisation (capital)	– General expression of societal wealth – Profit and interest-bearing capital – Qualitatively homogenous, quantitatively unlimited, divisible, mobile, reversible	– Measure and store of value, legal tender – Energy and matter side of valorisation are associated with qualitative heterogeneousness, quantitative limitations, indivisibility, locational uniqueness, irreversibility
Labour power and conditions of production	– Abstract labour as sole source of value and surplus value and substitutable condition of production – Land, raw materials fuels and other uncultivated resources are used as 'free gifts' from nature and sources of rents	– Labour power as bearer of concrete skills and specific knowledge – Reproduction of conditions of production as 'ficticious commodities' dependent on the (increasingly undermined and expensive) preservation of natural scarcities

Table 2.1 (Continued)

Form	Exchange-value moment	Use-value moment
Productive (industrial) capital	– Profit production especially through relative surplus – Contradiction of relative surplus production counteracted by the expansion of the scale of production (accumulation) – Increasing efficiency in the use of constant capital overcompensated by rising demands for natural resources	– Production of use values through the rearrangement of matter and energy – Expansion of production scale translates into increasing throughputs of raw materials, auxiliaries and so on especially as fossil fuels and advances exploitation of the natural conditions of production – Greater fossil fuel consumption disrupts the carbon cycle, triggering climate change
Accumulation and turnover of capital	– Valorisation logic forces entrepreneurs to reduce the non-productive stages of the capital cycle and to speed up turnover – Tendency towards overcoming distances in time and space or 'simultaneity'	– Production takes place under specific temporal and local conditions – Consumption of matter and energy is always linear and irreversible – Capital's 'expansionism' is accompanied by the degradation in the environment and increase in the greenhouse effect

Sources: Developed on the basis of Jessop (2002, p. 20), O'Connor (1998, p. 335), Burkett (2005, p. 144) and Altvater (1993, p. 201)

between the logic of unlimited valorisation, quantitative and geographic expansion, circularity and reversibility, which generally characterise the exchange-value moment of capitalist development and the natural laws that govern the work-process and the use-value aspect of production: qualitative matter and energy transformations from lower to higher entropy levels, and hence irreversibility, the narrowed stock of natural resources and, especially, their limited capability to serve as both sources and sinks for the permanently increasing flow and throughput of matter and energy arising from the exchange-value moment of production. Under capitalist auspices, it is not only exchange value and use value, but more generally the entire 'metabolic' relation between human beings and nature through the labour process that have the hallmarks of a 'rift' (Foster, 2000). Yet, due to the subjugation of labour process by valorisation process – and despite the fact that capitalism has transformed nature to a greater extent than any other mode of production – the recognition of the external, natural limits of production in general is severely complicated in capitalism and tends to take the form of rising costs on the supply side. It is ultimately due to the commodity form of labour products and to the corresponding separation of human producers from the means and objects of production that the economic system tends towards indifference with respect to its spatio-temporal and matter–energy specificities. The undermining of the material and ecological preconditions of the work process by the structural imperative towards valorisation and profit-making is expressed in the nullification and defiance of natural laws (Dietz and Wissen, 2009). While this led to the unleashing of productive forces, especially in the early stages of the capitalist mode of production, the latter is increasingly confronted with its external limits, which in the more recent stages of capitalist development express themselves in environmental imbalance and ecological disasters such as climate change.

3
The Regulation of Nature and Society in Different Capitalist Growth Strategies

Authors such as Burkett, Foster, O'Connor or Altvater have great merit for establishing that capitalist development is structurally bound up with tensions between the logics and the needs of the economic and of the ecological system. Elaborating on Marx's original critique of political economy, they all highlight the economic system's orientation towards unlimited and short-term valorisation, quantitative and geographic expansion, circularity and reversibility, while the principles that guide the ecological system involve stable and sustainable matter and energy transformations and throughputs – as well as irreversibility. Any long-term economic strategy that pays respects to the ecological system's guiding principles would further need to take seriously the fact that the earth's stocks of natural resources and their ability to serve as sources and sinks for waste from human production and consumption processes are limited. Though capitalist development cannot and does not get rid of the use-value element and of the material and energy side altogether, it nevertheless tends to negate and dispel them as much as possible. From the standpoint of individual capital, costs arising from the degradation of the environment are *faux frais* of production, which are, whenever possible, carried over to the general public – the taxpayer. The focus of the above-mentioned authors has, therefore, been on the paradox that capitalist production relations and productive forces tend to undermine and sometimes destroy their own social and ecological conditions of reproduction and, with them, the conditions for human life as such. Climate change is consequently seen as probably the most dangerous example, as it 'will inevitably destroy people, places, and profits, not to speak of other species life' (O'Connor, 1998, p. 166).

Thus capitalist development proceeds in not only socially, but also in ecologically contingent forms. Just as capitalist long-term development has to deal with social issues such as unemployment and social exclusion (Koch, 2006a, p. 5), humanity faces ecological disasters such as uncontrollable climate change if the logic of capitalist accumulation proceeds unrestricted. However, it is important to remember that the contradictions and tensions discussed in Chapter 2 are located at a relatively high level of abstraction: Marx's 'mode of production', where abstraction is made from institutional regulation and individual actors are reduced to their role of economic 'character mask', that is, to the roles they play in the production process.[1] Most recent Marxist ecological analyses remain at this level and fail to consider and compare actual capitalist societies at the level of what Althusser and Poulantzas called 'social formations'.[2] Though the analysis of the mode of production allows for insights into the general tensions between economy and ecology that characterise *all* capitalist societies, it does not sufficiently consider how these structural tensions are articulated in actual societies and in institutional circumstances. General tendencies such as the production of surplus population not only manifest themselves differently in different 'varieties of capitalism' (Hall and Soskice, 2001), but are also temporarily arrested under particular regulatory preconditions; this happens, for example, when the tendency towards a relatively 'redundant' population is overcompensated by an expansion in the scale of production and hence in absolute employment, as was the case during the postwar decades. With regard to climate change, the structural perspective allows for the important hypothesis that rising CO_2 emissions are caused by the capitalist mode of production's long-term trend of expansion of the scale of production and by the associated increase in material and energy throughput. But this perspective is too abstract and general to explain why CO_2 outputs per economic unit differ between Fordist and post-Fordist growth strategies, or why one capitalist country (say, Sweden) presents considerably lower CO_2 outputs per capita than another capitalist country (say, the USA). Since certain institutional peculiarities of a growth strategy will amplify the general tendency towards increasing CO_2 emissions, while others will modify or decelerate it, these points have to be considered systematically.

If we do not consider the fact that capitalism proceeds in different growth strategies and modes of regulation, then we run the risk of repeating the error of earlier generations of scholars, who thought that the social tensions and contradictions inherent in capitalism as a mode of production would lead to its inevitable (and in some cases immediate) collapse. And, just as capitalism proved capable of developing and coexisting with a regulatory network that – at least in the postwar decades – counteracted the otherwise dominant tendency of capital concentration, monopolisation and social polarisation, it would be premature to assume the end of capitalism on the mere grounds that fossil energy carriers will be exhausted, for example. Ecological limits to growth may turn out to be more flexible than a structural view based on the mode of production would suggest. New kinds of transnational institutional networks and agreements may arise to embed a new era of capitalist development, based on renewable energies. Dietz and Wissen (2009) and other recent German authors highlight capitalism's capability to adapt and adjust to changing 'external' conditions, including the foreseeable end of fossil fuels. Dietz and Wissen argue that biotechnologies are built upon genetic information and resources that are of crucial importance for the seed and pharmaceutical industries. Genetic resources crucially differ from fossil resources, since their productive use presupposes their maintenance and not their one-time productive consumption. Hence, a 'qualitatively new aspect in the area of biological diversity' is that 'large quantities of resources are often no longer required' (Görg and Brand, 2003, p. 267). This indicates that contemporary interaction processes between society and nature, which continue to be bound up with the imperatives of capital valorisation, are in some cases not built on the one-time consumption of matter and energy but on the continuous use of information and genetic resources. What is in dispute, hence, is not so much 'the use of "nature" as such, but the use for which form of production, under which social and ecological conditions and with which (ecological and social) consequences'. The governance of environmental issues such as biodiversity or climate change is 'not simply an economic process induced by market forces, but is something which is established politically. This means that ecological aspects become a factor in international competition, a strategic element of trade policy' (Görg and Brand, 2000, p. 391).

The recent debate in Germany indeed suggests that there is not just one type of capitalism but instead different, coexisting and competing ones, in which different capitalist growth strategies interact with nature within a social formation. As a corollary, the ecological destructivity that generally characterises capitalist development is also articulated in different ways and can be, in some institutional contexts, restrained. This involves the fact that, in particular industries, environmental protection is not in opposition with the profit motive but is one of its preconditions (Görg, 2003, p. 286), which leads to new divisions among the capitalist class. Dietz and Wissen (2009) discuss these points using the example of the overconsumption of raw materials and natural resources – which, for most entrepreneurs, is a means of valorisation, while for others it threatens profitability. When, to take an example by Dietz and Wissen (2009), the rainforest is cut down in order to use the wood for industrial purposes, then the very biological diversity is being diminished that produces the seeds that the pharmaceutical industries requires. Conversely, the declaration of specific natural territories as areas of nature protection for biotechnological use can come into conflict with the interests of oil and gas companies. Disputes between different capital factions arise as a result. The procedures and outcomes of these disputes contribute a great deal to the concrete way that nature is regulated in any given society and frames the adjustability of societies to environmental threats (Dietz and Wissen, 2009).

Another issue raised in the recent debate in Germany that has had direct impact on a social science understanding of climate change is that of an 'energy regime'. Altvater (2005, p. 72), for example, presupposes a systematic link between capitalism and fossil energy carriers, which he describes as a 'Trinitarian congruence' of capitalist forms, fossil energy carriers and European rationality. The converse argument would be that capitalism 'as we know it' will reach its limits in line with depletion of oil and other fossil energy carriers. Dietz and Wissen (2009) convincingly object that the assumed congruence between capitalist development and fossil fuel consumption underestimates the possibility of the emergence of a 'green capitalism', within which the global ecological and energy crisis might be, partially or totally, addressed. The two authors point to the case of DESERTEC, the building of an enormous solar–thermal power station in the Saharan desert.[3] If it goes into operation as planned, this station would continuously provide capitalist accumulation with the energy required – but this time on a solar basis and without, or with very few, greenhouse gas emissions. Whether or not such a green capitalism and a new energy regime on a renewable basis can emerge, or whether the capitalist mode of production itself is seriously threatened by the exhaustion of fossil energy carriers and by th aggravation of the climate crisis – as most recent Marxist analyses suggest – cannot be discussed and decided upon at the general structural level of Marx's *Capital*. In the last instance these are empirical questions, which need to be addressed within the context of actual capitalist growth strategies.

Just as unregulated capitalism is associated with an increase in inequality that undermines both system and social integration (Lockwood, 1992), and thus threatens the maintenance of the social order, it is also accompanied by ecological disasters on a hitherto unknown scale. Structural contradictions between economy and ecology must be kept within certain limits, so that the existence of the mode of production itself is not questioned. Tensions and contradictions take different forms, presenting themselves as a continuous development or as a rupture and corresponding to various forms of institutional regulation. In addressing elements of continuity and rupture in the relationship between capitalist development, nature and climate change, I will build upon the work of the regulation approach, which has been designed to consider the specific social, cultural and institutional forms and frameworks within which capitalist growth proceeds. The regulation approach – originally French – is not a unified framework; it represents instead a research programme or a set of different ideas that have been elaborated upon in the social sciences (Jessop, 1990; Koch, 2006a, pp. 19–24). The regulationists, at least in their 'Parisian' version, maintained the most important insights of Marx's critique of political economy, such as the differentiation between mode of production and social formation. Like Poulantzas (1975), the regulationists argued that the social formation cannot be simply 'derived' from concepts that correspond to the analytical level of the mode of production. However, they went beyond structuralism by stressing that capitalist accumulation depends on a range of social, cultural, political and institutional factors. Regulationists assume that capitalism 'develops through a series of ruptures in the continuous reproduction of social relations. Crises are resolved through an irreversible transformation which allows the fundamental or "determinate structure" of capitalism to continue' (Friedman 2000, p. 61). While the abstract features of capitalism are seen as largely transhistorical, both crises in the accumulation process and phases of expanded production are addressed in the context of their institutional embedding. The regulatory settings required for continued and expanded capital accumulation are socially, culturally and politically constructed and contested within a myriad of societal struggles, in which the relations both within and between social classes play a prominent role. Environmental struggles, including those related to climate change, are no exception to this rule.

The regulation approach

The regulation approach designed 'intermediary concepts', which expressed the largely non-variable conditions of the agents involved in the relations of production and exchange, as well as the historical changes these relations undergo during different phases of capitalist development. These concepts emphasised that the articulation of a given social formation in time and space corresponds to particular structural features and institutional forms. It was assumed that these forms are valid for a long period of time and make a crucial contribution to the stabilisation of the underlying structures of the mode of production. Regimes of accumulation are associated with certain historical phases and development paths, which are characterised by economic growth, 'under which (immanent) crisis tendencies are contained, mediated or at least postponed' (Tickel and Peck, 1995, p. 359). Such growth takes the form of compatible commodity streams of production and consumption, which are reproduced over a long period of time. Regimes of accumulation differ historically as to whether intensive forms dominated over extensive ones, export-oriented over import-oriented or vice versa, and as to whether the main focus of accumulation was on the production or the means of production or on the production of consumption goods (the two 'departments of production' in Marx's terminology). These regimes are further associated with a specific industrial paradigm, a dominant principle of the division of labour and a particular 'mode of consumption'. Consumption is not seen as an isolated phenomenon, the result of autonomous individual choices – as it is in neoclassical theory. Instead it is understood in its social genesis and context, or as 'an organized set of activities, which – while predominantly private – became subject to a general logic of the reconstitution of energies expended in social practices and the preservation of abilities and attitudes implied by the social relations' (Aglietta, 1987, p. 154). In other words, the regulation approach objects to the 'scholastic bias' in economics (Bourdieu, 2005, p. 7) or the tendency to construct increasingly abstract econometric models, 'which leads the scholar to project his thinking into the minds of the active agents and to see [his own representations] as underlying their practice'. The ascendancy of an aptitude for rational behaviour – with respect to both production processes and consumption practices – for example, is not an anthropological constant but instead the result of a long historical process during which it was inscribed in people's social and cognitive structures, practical patterns of thinking, perception and action (Koch, 2006b).

What and how much we buy and consume is of the greatest relevance for the carbon cycle, since such decisions are normally bound to matter and energy transformations, which more often than not necessitate the burning of fossil fuels. In contrast to neoclassical economics, the regulation approach insists that purchase decisions or the 'demand side' of economics are neither spontaneous nor 'individual' but are greatly influenced by structural factors such as income inequality and sales strategies. Instead, it agrees with sociological and anthropological research that point out that purchasing things is not in the first place about the goods themselves, but rather about the symbolic message that the act of purchase conveys. Both the acquisition and the possession of use values symbolise much of our social standing in society and of our identity and sense of belonging. Yet, if the rate of production of new, fashionable and desirable goods is high and accelerating, continuous efforts must be made by all social agents to reestablish or improve their original position and to distance themselves from other people. What Hirsch (1976) calls the competition for 'positional goods' is mediated through a genuinely social logic, which Bourdieu (1984a) refers to as 'distinction'.[4] But there is always the danger of vulgarisation, of devaluation through the emulation and generalisation of certain cultural practices, which once held the aura of legitimacy. The English saying 'keeping up with the Joneses' alludes to the imperative of continuously demonstrating one's unique taste and position in a society that always threatens to make one-time luxury goods accessible to all.[5] This sets in motion a never-ending cycle of definitions of 'taste' by the avant-garde and of keep-up strategies by the mainstream. This cycle plays into the hands of the valorisation interests of the various and reactive culture industries, but it contradicts the principal reproductive needs of the earth as an ecological system (Chapter 1).

A mode of regulation comprises an ensemble of social networks as well as rules, norms, and conventions, facilitating the seamless reproduction of an accumulation regime. The phrase 'mode of regulation' stresses the fact that capitalism does not reproduce itself only upon the basis of its immanent logic, but that its stabilisation requires institutional forms as well. These comprise five subdimensions: the wage relation or 'wage-labour nexus' (Bertrand, 2002); the enterprise form; the nature of money (Guttmann, 2002); the state; and international regimes (Aglietta, 2002). Capitalism also includes certain geographic scales, which determine the main spatial boundaries within which structural coherence is ensured (Brenner, 2004; Koch, 2008). The notion of 'spatio-temporal fixes' (Jessop, 2002) reflects the fact that different regulatory institutions deal with different issues not only by using different scales, but also over different time periods. Regulationists view the institutional forms that help to stabilise capitalist development during particular growth periods as the hard-won products of social struggles and of diverse and often contradictory interests. Hence, regulation in its concrete forms is not simply the product of the strategies of the dominant classes – which are themselves divided by different competitive interests and always reflect a degree of a compromise that includes concessions to the dominated classes and groups. Modes of regulation and patterns of governance vary considerably, depending on the nature of such compromises. Regulationists refer to a historical situation in which a regime of accumulation and a mode of regulation are sufficiently complementary to secure an extensive period of economic expansion and social cohesion by calling it a growth model or a model of development. The stability of such a growth model is further enhanced when shared values and norms help to bring about a commonsense value system that members of all social classes subscribe to. Regulationists such as Lipietz (1998) and Becker (2002) use Antonio Gramsci's notion of a hegemonic block with respect to the social classes and groups that adhere to these values and represent them (Gramsci, 1971). Where growth model and ideology correspond, authors such as Jessop speak of a successful mode of societalisation[6]. Such a correspondence frames and gives meaning to people's day-to-day beliefs and practices, thereby providing integration both at system and at social level.

In the recent debate in Germany, Christoph Görg and Ulrich Brand have applied the regulation approach to the society–nature relationship. Görg (2003, pp. 121–5), in particular, suggests that this relationship asserts itself through particular discourses, that is, through competing values and norms as well as through specific schemes of knowledge and interpretation, which are themselves linked to the power asymmetries between the different actors involved. In increasingly complex societies, there is not one but several competing ways of interpreting an environmental issue. What actually counts as 'environmentally relevant' is in fact variable over history and thus must be identified as an object of research in the context of changing societal integration and regulation patterns. 'Symbolic institutions' (Görg, 2003, p. 132) – such as institutionalised academia, the media and the political arena – all contribute towards a temporarily valid definition. Hence, the ways we see and understand 'nature' are socio-culturally mediated and subject to historical change. Societal power relations and the corresponding discourse patterns determine which ecological processes are perceived as 'problems' and deserve to be tackled. The question of whether climate change is occurring at all, whether it should be viewed as a serious issue and what could be done about it, for example, is part of the hegemonic struggles between opposing interest groups with different power resources. If capitalism can reproduce itself only through the establishment of a mode of regulation, which itself is contingent upon establishing institutional compromises in which different and opposing interests are temporarily balanced, then the analysis of the regulation of nature in general and of climate change in particular in specific capitalist growth strategies has to consider these institutional as well as the symbolic and ideological elements.

Consequently, Görg and Brand (2000, p. 374) regard 'the global ecological crisis' not as a crisis of the mode of production but 'primarily as an *institutional crisis* of the appropriation of nature by society'. The two authors interpret the current situation as one in which the definition of this 'crisis' is itself 'controversial and subject to varying interpretations', and in which different crisis explanations as well as ideas about new forms of an integration of socio-ecological issues and modes of societalisation coexist (Görg and Brand, 2000, p. 376). Social science research into concrete articulations of the integration of economic and environmental interests in particular spatial and temporal contexts must indeed carefully 'examine the interests involved, their unequally distributed capacities to assert themselves, and institutional structures' (Görg and Brand, 2000, p. 377). The very fact 'that international compromises are sought' (Görg and Brand, 2000, p. 377) indicates that contemporary growth strategies and their treatment of the environment do not simply 'assert themselves by the logic of capital' and are not simply implemented by powerful corporations and governments. Since growth strategies are instead subject to negotiation and concession, it is the task of the researcher to identify the historically particular institutional conditions in which such negotiations take place.

I agree with Görg and Brand that the general tension between capitalist development, ecological sustainability and climate change is expressed in different historical forms. Different accumulation regimes correspond not only to organised forms of capitalism, particular consumption norms and modes of societalisation, but also to specific forms of appropriation of nature by society. Subsequent chapters will, therefore, explore the issue of whether and to what extent different accumulation regimes are linked to specific 'energy regimes' – that is, the issue of the particular type and extent of the use of biotic, fossil and/or renewable energy carriers. Building upon regulation theory, I will examine the accumulation regime, consumption norm, regulatory system and energy regime of the two major capitalist growth strategies after the Second World War – Fordism and finance-driven capitalism. In other words, I will move on from the level of abstraction of Part 1, where the general tensions between capital accumulation and climate change have been discussed, and I will now include the particular institutional conditions in which conflicts on environmental issues such as climate change occur. In doing so, I will treat the question of whether or not capital valorisation can be temporarily brought into balance with the goal of a stabilisation of greenhouse gas emissions as an open and empirical one. Beginning with Fordism, I analyse how the development of a particular accumulation regime and mode of regulation is linked to climate change as a natural and social issue.

Part II
Fordism

"Fordism" was characterised – as an ideal type – by the parallel restructuring of both the technological and organisational basis of the production process and of the consumption patterns of wage-earners. Yet the historical development towards compatibility between both entities did not follow any 'grand plan', but was the rather accidental occurrence of largely independent socio-economic, political and cultural processes. Part II discusses this growth strategy in terms of its historical genesis and its geographic dissemination, with special emphasis on its material and energy basis. I argue that the environmental crisis and the specific issue of climate change became relevant within the context of the development of this 'new stage in the regulation of capitalism' (Aglietta, 1987, p. 117). Chapter 4 refers to the specific situation of the US in which Fordism emerged. Chapter 5 addresses the particular circumstances and ways in which this growth strategy spread to Western Europe and other parts of the world after the Second World War. Chapter 6 outlines the concomitant development of the mode of consumption and societalisation in the Atlantic space. Finally, Chapter 7 analyses the fossil fuel energy foundation of the Fordist economy and lifestyle.

4
The Origins of a New Accumulation Regime

Most economic historians agree that an extensive accumulation regime predominated in the second half of the nineteenth century. This regime was based on what Marx called the 'formal subsumption' of wage labourers and remained largely dependent on the subjective knowledge and skills of the workers. Only every fifth American worker was employed in a factory in 1900 (Robinson and Briggs, 1991, p. 622). The subsequent rise of factory production had little to do with its technological superiority over artisanal production (Gordon et al., 1982, p. 81); much more, it resulted from the strategy of augmenting output by increasing the overall number of workers. Rather than qualitatively transforming the work process, the basic business philosophy of the time indeed followed the motto 'so many hands, so much money' (Dawley, 1976, p. 28). Robinson and Briggs (1991, p. 652) demonstrate that this strategy was reaching the point of diminishing returns in 1880, 'as skilled workers in large factories were able to press for higher wages than their counterparts in small artisan shops'. Another factor that intensified competition between the largest factories was the increasingly closer link between product markets, from markets on a local scale to ones on a regional and eventually national scale, as a result of the quickly expanding railway system. Since accepting diminishing profit rates was not a realistic option, a solution was sought in the implementation of 'new machinery that could be operated by low-paid, unskilled labor and in new systems of labor control' (Robinson and Briggs, 1991, p. 653). By the turn of the century, the production methods applied had 'eliminated skilled workers, reduced required skills to the barest minimum, provided more and more regulation over the pace of production, and generated a spreading homogeneity in the work tasks and working conditions of industrial employees' (Gordon et al., 1982, p. 113). The basis of what Aglietta

(1987, p. 113) called the 'principle of mechanization' was the reversal of the relationship between labourers and the means of labour and the incorporation of the 'qualitative characteristics of those concrete labours previously performed by the dexterity of workers' into the machine system. Labour tended to become reduced to a 'cycle of repetitive moments that is characterized solely by its duration, the output norm'. Semi-automatic assembly-line production became the characteristic labour process of the new growth strategy. This production type was established especially for 'mass consumer goods produced in long production runs, and was subsequently extended upstream to the production of standardized intermediate components for the manufacture of these means of consumption' (p. 117). Product and process standardisation facilitated the vertical integration of production processes into both production and consumer goods industries.

Taylorism

The prolonged economic depression of the 1870s led to continually reduced demand, and thus unused industrial capacities. Manufacturers began to turn their attention away from technology to organisation, which, in turn, led to the 'scientific management movement in American industry' (Chandler, 1977, p. 272). Elaborating on the works of Henry R. Thorne, Frederick W. Halsey and Henry Metcalfe, Frederick W. Taylor (1947) pointed out the importance of standard time and output being determined 'scientifically', through detailed job analyses and time-and-motion studies of the work involved. He advocated a clear distinction between conception and execution, production and sales, marketing and finance and so on. Unlike in previous industrial paradigms, which had depended upon wage-earners' subjective abilities and skills, the entire work process was now designed to be emancipated from the specific qualifications of individual workers, and so it ensured that one worker could be quickly substituted by another. The function of workers was largely reduced to carrying out simple and repetitive tasks within the work process, while skills, qualifications and control of the work process were increasingly concentrated within the planning department. Jobs were broken down into their constituent parts, workers were timed by stopwatch and wage scales based on piece work were devised – so that a worker benefited from the expansion of output, but would receive less than the average wage and, indeed, would be forced to quit, if proved inefficient. By eliminating superfluous activity, the worker followed a machine-like routine and became more productive. Hence the individual worker lost all control over his or her working rhythm, and had to follow the continuous linear flow of the machine system; this systematically excluded subjective skills, human feeling and motivation, leaving many workers dissatisfied with their jobs. But, since job autonomy in the work process had been 'totally abolished', workers were normally 'unable to put up individual resistance to the imposition of the output norm' (Aglietta, 1987, pp. 118–19). Tasks could thus be simplified and standardised even further by fragmenting cycles of motion into the mere repetition of a few elementary movements. Workers were faced with an ever-increasing pace of work, combined with the curtailment of breaks, greater fatigue and nervous exhaustion – previously unknown – from which it was difficult to recover at the end of the working day. The by-products of the perfectioning of the new factory regime included a rising number of accidents on the assembly line, temporary and permanent disabilities, increases in the proportion of defective products and consequently in the time devoted to quality control, as well as an increase in absenteeism.

Taylor himself claimed that the essence of his management approach was not single elements such as piece work, task cards or time studies, but the fact that he initiated 'a complete mental revolution on both sides': management and labour. Applying his methods would indeed render ideas of class struggle old-fashioned. Instead, a mutual win-win situation would arise:

> both sides take their eyes off of the division of the surplus as the all-important matter, and together turn their attention toward increasing the size of the surplus until this surplus becomes so large [...] that there is ample room for a large increase in wages for the workman and an equally large increase in profits for the manufacturer.
>
> (Taylor, Testimony before the Special House Committee, cited in Maier, 1970, p. 32)

What the Taylorist system seemed to propose was indeed nothing less than the elimination of the problem of scarcity per se – no wonder that this proposition was taken up by a range of European intellectuals before and after the First World War. Among the first to introduce similar management principles was the Prussian bureaucracy under the leadership of von Stein, von Scharnhorst, von Gneisenau and von Moltke. These included centralism, detailed logistical planning, standardised operating procedures, the merit principle and the breakdown of tasks into their simplest components. The Prussian administrative system was subsequently widely emulated by other public and private organisations and became the model for Max Weber's ideal type of 'formal rational' action and 'bureaucratic' domination (Weber, 1978). The planning and control system were used by military commanders under Ludendorff to mobilise Germany's resources during the First World War. The 'economic war plan' (*Kriegswirtschaftsplan*) was subsequently adopted by the then nascent Soviet Union, which used it when designing GOSPLAN, its centralised planning system. In architecture, Gropius and Le Corbusier accepted Taylorist production methods and wished to limit their negative consequences on the workers' health and job satisfaction by designing a 'good factory aesthetic', which was to permit 'a more joyful cooperative effort' (Maier, 1970, p. 36). In politics, Fordist top-down planning inspired not only Russian Bolsheviks, but also movements as dissimilar as those of the German social democrats and the Italian fascists.

Though many of Taylor's ideas were incorporated into the evolving modern American factory organisation, industrial production was only rarely based on all of his principles. The original Taylorist system's weakness was thought to be 'its failure to pinpoint authority and responsibility for getting the departmental tasks done and for maintaining a steady flow of materials from one stage of the process to the next' (Chandler, 1977, p. 276). To provide overall coordination and control of throughput and, at the same time, to benefit from the functional specialisation proposed by Taylor, many major companies installed an explicit line and staff structure. While the operating 'departments or shops continued to be managed by foremen who were generalists and who were on the line of authority that came down from the president by way of the works manager or superintendent', the tasks of Taylor's planning department and functional foremen became those of a 'plant manager's staff' (p. 277). The Ford Motor Company was the first to apply systematically the principles of mass production in the early twentieth century. The compact vehicle named Model T, produced at Highland Park, Michigan, temporarily represented 60 per cent of the automobile output in the US (see Rae, 1969, p. 45). This spectacular success was due to the extremely low price of the Model T, which in turn was due to the great standardisation of components, manufacturing processes and, last but not least, product. Such standardisation required that parts were largely interchangeable. Ford achieved this by making use of advances in machine tools and gauging systems. These innovations, in turn, made possible the moving and continuous assembly-line system, in which each assembler performed a single, repetitive task.

As a result of the implementation of assembly-line production, labour productivity (as measured in working hours necessary to produce one car) increased threefold between 1909 and 1916 (Williams et al., 1992, p. 551). This productivity increase, in turn, permitted remarkable reductions in price, from $780 per car in 1910 to $360 in 1914 (Hounshell, 1984). Everything needed for the production of cars, from the raw materials on, was produced and organised by the Ford Motor Company. Vertical integration was preferred not only because of the perfectioning of mass-production techniques, but also because information processing was not well developed at the time. Accounting and finance were thus laborious and costly procedures. Direct, personal supervision seemed to be an efficient means of coordinating the flow of raw materials and components through the production process. Chandler (1977) describes the huge degree to which vertical integration and direct control were contingent on the fact that every staff category fitted into the company hierarchy and that staff carried out their allocated task. In order for every single worker to know his (rarely her) place, an entire supervisory army of middle managers and line managers, personnel specialists and so on had to be employed. In contrast to Taylor's original management theory, Ford rarely used time studies, since he 'favoured simpler more direct ways of driving the workplace' (Williams et al., 1992, p. 531). The authors also argue that the separation of execution from conception was not as rigid as might be expected on the basis of Taylor's writings, since suggestions about improvements from foremen or line workers were encouraged and sometimes led to promotion.

Social engineering or the embryonic form of a consumption norm

Fordist growth was due not only to improved machinery and the better design of factory work, but also to the revaluation of the working class as a source of demand. Henry Ford and his associates discovered that workers were useful not just as a source of profit production, but as a potential force of consumption of mass-produced commodities. When Highland Park opened, the company introduced a new wage system, which came to be known as the '$5 day'. This meant that Ford paid more than twice the normal wage for the industry of the time. At the same time, working hours were reduced from nine hours to eight, thereby converting the factory to a three-shift day. In contrast to earlier profit production on the basis of low wages, Ford began paying living wages that raised workers above subsistence level and made them potential customers for industrially produced commodities – and especially automobiles. The $5 day, hence, provided the embryonic form of a consumption norm for the working class (Koch et al., 2011). An important component of that norm was the implementation of a profit-sharing system. Yet being a satisfactory worker was not sufficient for participation; participation was contingent upon a particular lifestyle. Ford workers had to live according to a clearly defined code of behaviour (see Foster, 1988, p. 18) based on thrift, on having a home worthy of a Ford worker, on not sub-letting rooms in one's house to boarders, on not having an outside business of any kind, on not associating with – or allowing one's children to associate with – the 'wrong people', on cleanliness, on being married, on not drinking and smoking excessively, on prohibiting one's wife (in the case of a male worker) from working outside the home and on demonstrating progress in learning English.

The scheme was supervised by Ford's 'Sociological Department', which initially consisted of around 250 investigators. These visited workers' homes (and those of their neighbours and acquaintances) to determine their eligibility for the profit-sharing scheme. As a result of this monitoring process, initially 40 per cent of the Ford workforce failed to qualify (Marcus and Segal, 1989, p. 237). A further 28 per cent were disqualified from the scheme for one reason or another during the first two years. The Sociological Department revised the suitability criteria every six months. Those who initially had not qualified could attempt to change their lifestyle during the intervening period, but continued failure led to dismissal. Furthermore, the Sociological Department provided Ford employees with lessons in family and household budgeting, home management and hygiene and cost-effective shopping. In addition, the 'Ford English School' instructed non-English speaking employees in the virtues of efficiency, thrift, industry and economy, and provided compulsory courses in domestic, community and industrial relations. According to Marcus and Segal (1989, p. 237), it was the school's mission to weld systematically the 'diverse groups compromising Ford's labor force into a standardized, dependable cohort'. Therefore, the new consumption norm went much further than creating demand for mass-produced cars via the $5 day. It was equally an instrument of human and social engineering, since it extended the power of the employer over employees from the production process to the sphere of leisure and consumption. It crucially structured 'the conditions under which labor power was reproduced within the home' (Foster, 1988, p. 123). For those workers who worked and lived by the new rules, participation in the profit scheme and, on this basis, the possession of an automobile and their own home became status symbols of upward mobility and of leading an approved lifestyle. For those who did not do this, the acquisition of goods like a car remained a distant dream. These workers were castigated for lifestyles that deviated from the consumption norm (Koch et al., 2011).

Regulationist researchers emphasise that the development of compatible production and consumption norms normally does not proceed in planned ways, but rather through accidental occurrences of relatively autonomous entities (Lipietz, 1998). Sorensen (1956, pp. 137–41) confirms this evaluation by using the example of the $5 day. Having worked for and with Ford for 40 years, he outlines that the $5 day was not the result of Ford's wish to share profits with his employees. In contrast – Sorensen reports from actual company meetings where the policy was discussed – the new wage policy was a response to extremely high labour-turnover rates, which had reached 370 per cent by 1913 and which had forced Ford to spend nearly $2 million training new workers every year (Marcus and Segal, 1989, p. 236). Leaving Ford was indeed the only resistance option available for an individual worker when faced with the extreme physical and mental burden associated with being a mere appendage of the mechanical system. Ford himself took great pride in the fact that his company was 'not unionized', and that his 'labour control regime was designed to prevent informal resistance' (Williams et al., 1992, pp. 532–3). Williams et al. further note that there was an external labour market at the time, so that leaving Ford was a feasible and popular option. The introduction of the $5 day was, therefore, not so much a humanitarian act – although Henry Ford was indeed viewed as a 'national hero' by many – but a counter-strategy designed to stop high labour turnover. Ford himself (quoted in Marcus and Segal, 1989, p. 236) was in no doubt about the economic nature of this decision and maintained that the $5 day constituted 'efficiency engineering' in the first place and would stand as 'one of the finest cost-cutting moves we ever made'. And indeed, labour turnover fell to 16 per cent in 1915 (p. 237).

Limits to standardisation

The limits of the extreme standardisation of procedures and products became obvious as early as the mid-1920s, with the decline of the Ford Motor Company and the simultaneous rise of General Motors (GM). Ford's product standardisation had gone so far that the Model T was only available in black. GM's Alfred P. Sloan, Jr, capitalised on customer discontent. He conceived of post-1920 car purchasers as primarily second-time buyers and contended that, as a group, their objectives differed from first-time owners. To stimulate this new demand, the 'General Motors Acceptance Corporation' provided loans at low interest so that people could buy the higher-priced GM cars. Sloan also fostered the creation of a new second-hand car market, which made it more attractive for motorists to sell cars in order to purchase higher-priced GM cars. On the basis of technological innovations such as 'high-compression engines fuelled by tetraethyl lead to prevent knock [...], independent front-wheel suspensions to promote handling, and automatic transmissions' (Marcus and Segal, 1989, pp. 284–5), GM provided five different cars in a continuous line of products, oriented at different customer groups, from the poorest American able to afford a car to the wealthiest. By 1922, each of the five vehicles was sold in a different price range and was equipped with a range of different features (p. 285). Sloan recognised that each of the five automotive divisions needed its own staff (executive, engineering, design, production, marketing and sales) and he gave divisions an autonomy normally associated with independent companies; he established interdivisional groups to harmonise activities and smooth out difficulties and gave ultimate authority to the corporation's chief executive. Decentralisation and product diversification culminated in the annual model change concept in 1925. Sloan introduced one basic GM style each year, but left it to each division's designers to modify the form to suit target groups. For example, in 1927 Cadillac presented buyers with 500 colour and upholstery options and 'reconciled mass production with product variety by using some car parts in more than one division' (p. 286). As a result, GM captured the American US automobile market and Ford had to respond by abandoning his Model T in 1927 and by introducing his own annual model change in 1933. The company switched to flexible mass production and embraced the concept of a product line aimed at a wide spectrum of buyers.

In summary, in the US, mass production began to develop as an industrial paradigm in the early twentieth century. During the first three decades of the century the system was refined towards flexible mass production due to technological innovations and in order to find new customers for durable goods such as automobiles and household appliances. However, despite early attempts at introducing individual loans and consumer credit, the increased output of consumer durables on the basis of the new production methods and technologies was not matched by the same level of American purchasing power: 'In the 1920s earnings did not keep up with increasing productivity, so that by 1929 available markets were incapable of absorbing the full-employment output of American industry' (Agnew, 1987, p. 66). When firms began to shed workers as a result of the relative overproduction of durable goods, workers were unable to keep up with the credit payments they had started when employed. The overproduction crisis was extended to the financial system and to the stock market. Further elements of the ensuring general economic crisis, which came to be known as the Great Depression, were a dramatic downturn in the construction industry and a downward trend in world agricultural production (ibid.). During the following period, 1929–1933, the economy collapsed; the gross national product (GNP), adjusted to current prices, declined by 46 per cent, industrial production fell by more than half, wholesale prices fell by one-third and consumer prices by one quarter (p. 65). As a result of the crisis, employment decreased by nearly 20 per cent and unemployment rose from 1.5 million to roughly 13 million. Unlike previous crises, where there had been relatively quick recoveries, this one seemed to be endless and was resolved only by the advent of the Second World War.

5
The Geographic Extension of Fordism

From a regulation theoretical perspective, the Great Depression can be understood as the result of a lack of compatibility between the new production methods and an inadequate mode of regulation, which did not enable wage-earners to increase their consumption sufficiently to match rapidly growing industrial output. It was the regulation aspect of the growth strategy that had to change. During capitalism's greatest crisis to that point, President Roosevelt, in power from 1933 to 1945, initiated a 'New Deal' of socio-economic regulation. The cornerstone of the new strategy was the provision of a minimum standard of welfare through economic stabilisation and social policies. This implied a reinterpretation of the role of the state in socio-economic affairs. Once the state was no longer exclusively regarded as an impartial 'watchdog' agency, more and more areas were influenced by the New Deal – even once sacrosanct domains, such as prices and the valuation of money. According to Agnew, the New Deal did not end the Great Depression but made 'life bearable or even possible for large numbers of people', and 'it certainly headed off dissent'. But the probably most lasting effect was that 'it legitimized the idea of a strong federal government usually in partnership with, rather than opposed to, big business' (Agnew, 1987, p. 69). A new corporate coalition between American government and business evolved and became the structural basis for US strategies in the domestic and global economy over the following decades. Yet, whatever the immediate effect of the New Deal, it was not until the US entered the Second World War that the American economy was lifted out of stagnation. Curing depression and stabilising the economy through the rapid increase in federal expenditure during the war, which generated more than one-third of the gross national product (GNP), was a new form of crisis management. More than half of total war expenses were financed through public borrowing, so that national debt increased astronomically. Yet under wartime conditions, this posed few problems, especially in relation to inflation. Since the debt was internally held, both individuals and banks acquired government securities to offset their liabilities. Furthermore, individual and business assets grew faster than the rate of indebtedness thanks to the rapid economic growth produced by deficit financing. The wartime experience of ending the depression seemed to justify the continuation of the 'grand coalition' of government and business to prevent the recurrence of crises and to maintain economic growth: 'Growth at home and expansion abroad were such clear possibilities that they unified the interests of previously contending groups in such a way that the nature of American political thinking underwent a major transformation in the 1950s' (p. 71).

In the specific postwar circumstances of the 1940s and 1950s, US firms expanded towards Western Europe in order to circumvent trade barriers and to take advantage of labour cost differentials.[1] Investment in the recovering European markets was attractive for American business as these markets could be captured relatively easily given the weak competition. Particular work processes that had become increasingly expensive in American locations were outsourced because of wage costs, taxes and so on. US governments encouraged and assisted this overseas expansion of capital. According to Agnew (ibid.), both American business and foreign policy were designed in the 'belief in the need to export the American model of business–government cooperation. The free movement of capital, free trade and anti-Communism combined as the major elements in American foreign policy' (p. 76). Indeed, the Marshall Plan, in place since 1949, cannot be fully understood without considering the general 'containment' orientation of US foreign policy in relation to the Soviet Union and its Eastern European allies. While US military forces opposed and threatened those countries that followed a communist development road (or were forced to do so due to Soviet pressure), generous economic support rewarded those countries that opted for market-oriented development options. The Marshall Plan, hence, not only helped to rebuild the destroyed Western European economies, but it also created enormous marketing opportunities for the US industry, thereby helping it to convert from a war to a peace economy, and boosted the methods of efficiency, productivity and growth that had been tried and tested in the US prior to the war but had run up against the limits of insufficient domestic purchasing power, especially on the part of wage-earners. For the Western European societies, whose economies had hit rock bottom, technological auxiliary programmes and the import of American machines and skills became the economic base for their modernisation. Then arose what was later called 'Atlantic Fordism' (Jessop, 2002).

The Atlantic space

One favourable condition of the postwar period was the fact that producers could count on a quasi-'infinite' demand for mass-produced goods such as automobiles and for household appliances such as televisions and washing machines. Unlike during the 1930s, when solvent consumers were scarce, during the era of the postwar reconstruction there was stable and expanding demand for both consumer goods and the means of production to build them. Agnew (1987, p. 76) confirms this comparison by empirically showing that 'much of the growth in the US economy in the 1950s and 1960s depended upon increasing sales of consumer durables'. While this had also been the main driver of economic growth in the 1920s, 'the difference was that the 1950s was a time of expanding markets and relatively free trade for American business whereas the 1920s was the reverse'. Since most Western European households did not yet own durable goods such as household appliances, mass production could become the technological basis for their speedy generalisation. The turnover of fixed capital was accelerated by the continuing increase in the number of products, which reduced the costs of one single product. Profits were supported by the consumer demand, which in turn was based on increasing real wages, and these were usually determined by collective agreements and tied to expected growth in productivity. The ability of the (mostly) unskilled workers to achieve competitive wages ultimately depended on their political organisation, especially in trade unions. Another important difference from the prewar period (especially from Henry Ford's trade-union free factories) was that organised labour was now a part of the corporatist arrangement, which took the form of a triangle of government, business and trade unions. This was supported by the fact that, by the 1950s, increasing mass production had made the working class the most sizeable group in the social structure of advanced capitalist countries. As wage labourers who were already 'organised' by their employers at company level, they could be mobilised in trade unions relatively easily. Their rising power was not only reflected in company-level co-determiniation, but was also increasingly recognised in the wider society, leading to forms of centralised collective bargaining either at industrial or at national level in most Western European countries (Boyer, 2002; Koch, 2005).

Table 5.1 Rates of GDP growth per capita in the Western world: 1820–1989

	1820–70	1870–1913	1913–50	1950–73	1973–89
Austria	0.6	1.5	0.2	4.9	2.3
Belgium	1.4	1.0	0.7	3.5	2.0
Denmark	0.9	1.6	1.5	3.1	1.7
Finland	0.8	1.4	1.9	4.3	2.8
France	0.8	1.3	1.1	4.0	1.9
Germany	0.7	1.6	0.7	5.0	1.9
Italy	0.4	1.3	0.8	5.0	2.6
Netherlands	0.9	1.0	1.1	3.4	1.3
Norway	0.7	1.3	2.1	3.2	3.1
Sweden	0.7	1.5	2.1	3.1	1.7
UK	1.2	1.0	0.8	2.5	1.9
Ireland	–	–	0.7	3.1	2.9
Portugal	–	0.3	1.4	5.6	1.7
Spain	0.6	1.4	0.2	5.1	1.8
Australia	1.9	0.9	0.7	2.4	1.7
Canada	–	2.3	1.5	2.9	2.4
US	1.2	1.8	1.6	2.2	1.6

Source: Maddison (1995, p. 97)

Table 5.1 shows the extraordinary dynamic of Fordist economic growth. In Western Europe and in the Atlantic space in general, the implementation of the new growth strategy resulted in fast growth rates of gross domestic product (GDP) and productivity during the 1950–1973 period (Table 5.2). The growth rates of what came to be known as the 'golden age' are especially spectacular when compared to those of pre- and post-Fordist periods. Prior to 1950, GDP growth rates of above 2 per cent were very rare (the exceptions being Canada in the period 1870–1913 and the two Nordic countries Norway and Sweden in the period 1913–1950). Between 1950 and 1973, however, the GDP grew much faster than ever before, with rates between 3 and 5.6 per cent in most countries. In relation to the UK – the main European exception, which displayed an average GDP growth of only 2.5 per cent – earlier comparative research suggests that technologies and forms of work organisations associated with Taylorism were never implemented to the same extent in the UK as in other European countries. This was not least due to the British labour movement, which defended specific divisions within the working class according to skills, tasks and job rules (Koch, 2006a, p. 148; Tickel and Peck, 1995, p. 362). The relatively small GDP growth rates of the US must be understood against the

Table 5.2 Comparative levels of labour productivity: 1913–1989 (US GDP per labour hour = 100)

	1913	1950	1973	1989
Austria	48	27	59	75
Belgium	61	42	64	89
Denmark	58	43	63	66
Finland	33	31	57	70
France	48	40	70	95
Germany	50	30	64	79
Italy	37	31	64	81
Netherlands	69	46	77	92
Norway	43	43	64	83
Sweden	44	49	76	81
UK	78	57	67	81
Australia	93	67	70	78
Canada	75	75	83	91
US	100	100	100	100

Source: Maddison (1995, p. 123)

background of the already high level of the GDP in 1950. This is also reflected in the superiority of US labour productivity. Table 5.2 comparatively measures the development of labour productivity in selected Western countries, using the US as a yardstick. In 1913, the UK had already lost its advanced position in labour productivity, which it had held throughout the nineteenth century. Even so, in 1950, labour in the Western European countries was less than half as productive as in the US. However, with the generalisation of Fordist production methods over the subsequent decades, these gaps in productivity became smaller and smaller. By 1973, labour productivity in Western Europe was already between 59 per cent (Austria) and 76 per cent (Sweden) of the US level. Gaps in productivity continued to shrink during the period of 1973–1989.

In addition to the Marshall Plan aid and the import of US technology, new forms of transnational cooperation made possible the process of 'catching up' in Europe. In Western Europe, the European Community of Steel and Coal (ECSC) and, later, from 1958 on, the European Economic Community (EEC) supported the emerging national growth projects above all through a common agricultural policy. By granting subsidies, setting production quotas and providing protectionist external tariffs, ECSC and EEC supported the global competitive position of the European agricultural sector. Increases in the productivity of the agricultural sector provided the necessary labour power for the rapidly expanding industrial sector (Ziltener, 2000). Early steps towards European integration consequently included the promotion and protection of agricultural production as well as the development and application of a joint trade policy. The agricultural sector served as a model for further attempts to attain a greater degree of integration in other sectors (Heeg and Oßenbrügge, 2002). The increase in the consumption of energy resources was another crucial issue, and it was dealt with in the first European treaties concerning mining, the steel industry and the use of nuclear energy. Fast growth rates in GDP and productivity allowed for rising real wages and for employment to move towards full capacity. High levels of domestic demand promoted full employment and led to an unparalleled investment boom in Western Europe.[2] Wages were indexed to productivity growth, while wage and credit policies were directed towards the creation of effective aggregate demand in national economies. This required strong governments, which helped to secure this demand by means of policies designed to integrate the circuits of the capital and consumer goods industries, and by mediating the conflicts between capital and labour over the individual and social wages. In their *welfare* role, governments crucially contributed to the decommodification (Esping-Andersen, 1990) of labour power by using the growing income from taxation for substantial income redistribution and the introduction and/or expansion of welfare systems – which, in turn, provided a guaranteed minimum standard of living for those who did not participate in the employment system.

Even though the national level was the main spatial target of socio-economic regulation (Brenner, 2004; Koch, 2008), Fordism, in the US and abroad, would have been unsustainable without international regulation. The elites in the US and in other leading capitalist countries wished to avoid a crisis of the dimension of 1929. The breakdown of the system of coupling national currencies to a national economy's available gold reserves was seen as an important contributory factor to the Great Depression. It was not until 1944, with the introduction of the International Monetary Fund (IMF) and the World Bank (WB) at Bretton Woods, that the actual generalisation of Fordism began. This was the first time that institutions were created to regulate not only the national, but also the global market. The General Agreement on Tariffs and Trade (GATT) in 1947 and the creation of common regional economic spaces were further expressions of this development, which led to the gradual reduction of national tariff and trade barriers. Subsequently, national central banks had a much greater degree of autonomy in their monetary policies. Crucial economic decisions on parameters such as the convertibility of national currencies or the determination of whether, and to what extent, a trade partner was creditworthy were not taken more or less automatically, on the basis of the gold standard, but increasingly via central banks and national governments. The prime role of the World Bank and of the IMF was to determine a system of fixed rates of exchange, within which both institutions could give loans to nation-states, for example in order to compensate for deficits in the balance of payments. Thus, differences in economic development between different states could be accommodated and sometimes reduced (Hall and Midgley, 2004).

By continuing to tie its own currency to the gold standard, the US supported this international regulation of money and loan transfers. The dollar functioned as 'global currency', and thus it was possible to compensate for deficits in the balance of payments, which accompanied the foreign trade surplus of the US after the Second World War. The US served as the reserve-currency country for the capitalist world economy. Through its role as a monetary overseer, the US was able to finance its deficits – the 'costs of *Pax Americana*', in Agnew's formulation. The US did not run deficits throughout the 1960s and provided loans to others 'as a by-product of pursuing domestic growth and overseas expansion at the same time' (Agnew, 1987, p. 76). As Agnew outlines, the system of fixed exchange rates pegged to the dollar was a 'major boost to takeovers of foreign industries by American business', since each nation was responsible for keeping the value of its currency within 1 per cent of its set value. To keep within that range, central banks were required to sell or purchase their own currency on foreign-exchange markets. Hence, by running large deficits, the US effectively 'forced foreign central banks to buy excess dollars with their own currencies' in order to decrease the supply of dollars in global circulation. This provided American investors with the foreign currencies necessary to buy assets in France, Italy, Britain or Germany. At the price of international 'monetary stability', foreign central banks were put in the position of 'financing the takeovers of their own industries' (p. 85).

Developing countries and 'real-existing' socialism

Development loans from the IMF and the WB were at the same time an important financial basis for attempts to industrialise developing countries. Some countries attempted to achieve a delayed Fordist development through policies of industrial import substitution, in which the national state played the key role. The state supported the industrial-isation processes through measures ranging from high tariffs designed to protect the domestic market and cheap loans and tax advantages for local investors to direct state investments in employment and infrastruc-ture. In the case of larger countries like Brazil, Mexico or South Africa, authors such as Feldbauer et al. (1995) talk of a 'peripheral Fordism'. This delayed industrialisation was contingent on the accessibility of for-eign loans at interest rates that could be paid back from the profits of the sales of commodities produced in factories, which were originally built thanks to the loans. Import substitution could work only up to the point where the profit rates from the sale of these commodities were above the interest rates that had to be paid back to the international banks, the IMF and the WB especially. Table 5.3 shows that import substitution

Table 5.3   Rates of GDP growth per capita in the developing world: 1820–1989

	1820–70	1870–1913	1913–50	1950–73	1973–89
South America	0.3	1.1	1.4	2.5	0.6
Argentina	–	1.9	0.7	2.1	–1.2
Brazil	0.2	0.3	2.0	3.8	1.7
Chile	–	–	1.7	1.2	1.5
Colombia	–	–	1.5	2.1	1.8
Mexico	0.4	1.1	1.0	3.1	1.0
Peru	–	–	1.4	2.5	–1.2
Asia	0.1	0.6	–0.1	3.5	4.2
Bangladesh	–	–	–0.3	–0.7	2.2
China	0.0	0.3	–0.5	3.7	5.7
India	0.0	0.3	–0.3	1.6	2.7
Indonesia	0.2	0.5	–0.2	2.1	3.4
Japan	0.1	1.4	0.9	8.0	3.0
Korea	–	–	–0.2	5.2	6.4
Pakistan	–	–	–0.3	1.8	2.8
Taiwan	–	–	0.4	6.2	6.1
Thailand	–	0.4	0.0	3.2	5.2
Africa	–	–	1.2	1.9	–0.3
Ivory Coast	–	–	–	2.9	–1.2
Ghana	–	–	1.1	–0.1	1.4
Kenya	–	–	–	2.6	0.7
Morocco	–	–	–	2.4	–1.5
Nigeria	–	–	1.2	2.3	0.2
South Africa	–	–	1.2	2.3	0.2
Tanzania	–	–	–	2.4	–1.4

Source: Maddison (1995, p. 97)

strategies were not unsuccessful during the growth period of Fordism. With the exception of Asia, which even managed to increase economic growth during the period 1973–1989, GDP growth rates among developing countries were considerably higher during the period 1950–1973 than before or after.

Finally, and under different circumstances, the 'real-existing' socialist world was equally structured by the characteristics and imperatives of Fordist growth. In 1910, Rudolf Hilferding (1981) had already described the era of 'finance capital' as one of cartelisation and concentration of capital. For him, the corresponding high degree of monopolisation of economic wealth, in combination with a greater role of the state, pointed towards a transformation within the capitalist economy of the time – a transformation that would help to stabilise it in the short term but would also facilitate the ultimate economic transition towards socialism. Lenin accepted this view and endorsed a stage of 'state capitalism' on the long road to socialism. The German war economy, with its great emphasis on central planning, provided some orientation in relation to the economic aspect of this transformation. But, above all, Lenin openly welcomed Taylorism as a means of not only optimising economic output, but also of guaranteeing Soviet power. On the one hand, he scathingly described the Taylorist system as 'a combination of the subtle brutality of bourgeois exploitation'; yet, on the other hand, he praised the 'greatest scientific achievements [to date] in the field of analysing mechanical motions during work, the elimination of superfluous and awkward motions, the working out of correct methods of work, the introduction of the best system of accounting and control' (Lenin, 1947, p. 327). He considered it a foremost priority for the nascent Soviet Union 'at all costs [to] adopt all that is valuable in the achievements of science and technology in this field', culminating in the postulation that 'we must organize in Russia the study and teaching of the Taylor system and systematically try it out and adapt it to our purposes' (ibid.). Thomas Hughes (2004) describes in great detail how the Soviet Union acted on Lenin's suggestion. The embracement of Fordism and Taylorism went so far that American experts were imported and American engineering firms were commissioned to build parts of the new industrial infrastructure. Hughes directly traces the Five-Year Plan in particular and the centrally planned economy in general to the influence of Taylorism. It was only as the Soviet Union developed and grew in power that both sides – the Soviets and the Americans – chose to ignore or deny the contribution of American ideas and expertise. The Soviets did this because they wished to portray themselves as creators of their own industry,

Table 5.4   Rates of GDP growth per capita in Hungary, Czechoslovakia and the former USSR: 1820–1989

	1820–70	1870–1913	1913–50	1950–73	1973–89
Czechoslovakia	0.6	1.4	1.4	3.1	1.3
Hungary	–	1.2	1.2	3.5	1.7
Russia/Former USSR	–	0.8	2.3	3.6	1.0

Source: Maddison (1995, p. 97)

and not indebted to their rivals. The Americans did the same because they did not wish to acknowledge their part in creating a powerful rival. Once the system competition and the East–West opposition had been established, the slogan in the USSR and in smaller satellite states (such as the German Democratic Republic) was to 'overtake the West without catching up'. And indeed, as long as Western growth was largely based on mass production, it could be emulated relatively easily in centrally planned economies. It was only from the second part of the 1970s, when Western economic units became smaller and more decentralised, that the real-existing socialist economies proved insufficiently flexible to deal with changes and increasing complexities.

However, 'state-socialist Fordism' was not unsuccessful at all during the postwar period. As in the Western and developing countries, economic growth rates per capita in the period 1950–1973 were significantly above those before and after this period (Table 5.4). When comparing long-term economic data for three countries of the former socialist block with those of Western countries, one should consider that these economies, despite being almost completely destroyed under the German occupation, had not qualified for Marshall Plan aid or for any import of US technology. Since these types of support were reserved for countries that opted for capitalist development strategies, economic growth rates of over 3 per cent in the period 1950–1973 were a remarkable achievement indeed.

6
Mode of Societalisation and Consumption Norm

The productivity growth associated with the achievement of economies of scale was a prerequisite for the simultaneous and proportionate development of the two departments of social production, namely production and consumption goods. The percentage of wages to total employers' costs decreased (or, in Marx's terms, the 'organic composition' of capital grew), but the real wages of workers also increased. Employment was able to grow since the total volume of capital rose by a greater proportion than the increase in the number of workers made redundant due to productivity gains in the work process. The cheapening of industrial products raised the purchasing power of wage labourers, so that both the employers' profits and the employees' real wages increased. The state benefited from this favourable situation and used its growing income from taxation for the expansion of a welfare state system, which, in turn, guaranteed a minimum standard of living for those who did not participate in the labour market. Not only was the working class actively integrated into the growth project of Fordism, but also, for the first time, the unemployed and the recipients of welfare entitlements, the pensioners and the students (in some countries) received independent incomes, which the state raised via taxation and subsequently redistributed to these groups (Koch, 2001). However, social inclusion remained incomplete, since women were partially or completely excluded from economic activity. Until the 1970s, most European countries, with the exception of the Nordic countries, were characterised by an almost full capacity utilisation of the male workforce; for most women, in contrast, gainful employment was only acceptable before marriage and motherhood. Non-mainstream forms of family life were not seen as respectable and usually were not even mentioned in public or social science discourse.

Since women's income remained largely dependent on the outcomes of informal exchange relations with the 'male breadwinner', Fordism confirmed and reinforced the basic structures of the patriarchal division of labour.

The growth of Fordism left a deep impression on the social structure of advanced capitalist societies. The following features stand out among the structural features of the growth period of Fordism: the rising percentage of dependent employment in general and of industrial employment in particular; full employment and full-time jobs; but also, due to the predominant male-breadwinner model, a low percentage of female economic activity. With the continuing generalisation of fully or semi-standardised work processes and their spatial concentration, a relatively homogeneous working class emerged that learned to defend its interests confidently. At the same time, process and product standardisation were accommodated by an expansion in the size of the units of production, and this coincided with a concentration of capital in fewer and fewer companies and a diminishing level of self-employment in the economically active population. Yet the changes in living conditions and lifestyles associated with the upswing of Fordism went far beyond those that are measurable in quantitative statistical operations.[1] Many societal areas were affected, not least the conditions for the politicisation of collectives such as social class. During the pre-Fordist period, the power and influence of the working class were due, above all, to traditional ways of life and normative orientations that often contrasted with the upcoming industrialism, with its new technologies and lifestyles – a contradiction that found its expression in anti-capitalist leanings such as those exercised by the Luddite movement. Many of these conflicts were due to the fact that the labour power was merely formally subsumed by capital and its reproduction was largely mediated through non-capitalist channels. This situation changed fundamentally with the introduction and generalisation of the Fordist growth strategy, which qualitatively adapted the conditions under which the worker's individual reproduction took place and 'cut off' the 'ties of family, or bonds of neighourhood proximity or supplementary activity' that had linked them to a 'non-capitalist environment' (Aglietta, 1987, pp. 153–4). With the enormous reduction in the price of essential consumer goods, the condition for what Burghart Lutz (1989) called 'internal takeover' (*innere Landnahme*) of ever more spheres of life by capitalism became established: this was the participation of the 'masses' in commodified forms of consumption that until then had been monopolised by the dominant classes.

Though spatially and temporally separated, the spheres of work and non-work were nevertheless closely linked, in novel ways. A rather obvious link was the fact that recovery from the increased exhaustion of labour power created by Taylorised work processes had to take place during a limited period of time each day, and in one place – the home. This recovery was complicated by the additional temporal constraint of having to travel to the workplace, which was often not necessary in pre-Fordist periods. Links between the workers' roles in the production process and the sphere of consumption were further amplified by the ideal of 'standardised housing' as the privileged site of individual consumption and of the automobile as the means of transport compatible with the separation of home and workplace. Whilst remaining commodities for private use, these are also durable goods, whose acquisition exceeds the purchasing power of current wages. Taking up mortgages and consumer loans made the wage labourer's existence further dependent on continuing participation in the work process and made rebellious behaviour against the dominant production and consumption model increasingly difficult to afford. Another factor that helped to compatibilise the accumulation regime and the consumption norm was that many services and activities that used to be carried out at home were now commercialised and bought and sold on the market (Altvater, 1992). For example, the use of dishwashers and washing machines replaced washing by hand, and vegetables were bought in shops and supermarkets instead of being grown in kitchen gardens. Frozen 'ready-to-serve' meals often substituted home-cooked meals (Hirsch and Roth, 1986; Altvater, 1992). At the same time, the relative importance of personal services was reduced in the overall employment system and became a recruiting ground for industrial employment. To the same extent that services and products took the form of commodities, the percentage of wage labour within all types of economic activity increased: dependent employment and the corresponding income increasingly became a requirement for the acquisition of indispensable goods and services.

Suburbanisation and the rise of individual mobility

The more the spheres of production and consumption become linked, the more difficult it is to view individual acts of purchasing goods as 'individual choices'. Two areas concerning the 'choices' people made in their private live came to be particularly affected by the new consumption norm: housing; and the rise of a suburban lifestyle based on individual mobility. Standardised housing 'put an end to unhygienic and unsafe interiours', which had characterised pre-Fordist housing among the working class, and 'permitted the installation of household appliances that saved domestic labour' (Aglietta, 1987, p. 160). It became also a symbol of status, since housing could be bought rather than simply rented. Mass-produced housing via prefabrication techniques reduced costs to a point where, extended over the overall term of payment, it was, as Aglietta observed, less of a 'burden on the working-class wage of the 1950s than were the rents extorted by landlords of the inter-war years' (Aglietta, 1987, p. 160). Yet the growing material welfare and rationalisation in house building brought with it a parcelisation of households, so that traditional 'proletarian' working class milieus were gradually replaced by a homeowner culture. The new dwellings were either flats or, in the case of the upwardly mobile, semi-detached or detached family homes, with functionally differentiated 'departments' for eating, sleeping and children (Hirsch and Roth, 1986). Using the example of the French housing market during the heyday of Fordism, Bourdieu empirically demonstrates that economic choices[2] such as whether to buy or to rent or whether to buy an old house or a new one are far from 'individual', or even 'natural', but depend instead on the socially constituted economic dispositions of the agent (the 'demand' side) and on the supply of dwellings. Contrary to neoclassical economic theory, which treats both supply and demand as 'unconditioned givens', Bourdieu (2005, p. 15) argues that they are both shaped, more or less directly, by state socio-economic policies, and especially by housing policies. It is the latter that determine the conditions for owning, renting, constructing and so on and define regulation for taxation and the qualitative standards of housing. In short, the 'state', especially during its strong, interventionist and almost creative form of the 1950s and 1960s, 'contributes very substantially to producing the state of the housing market' (Bourdieu, 2005, p. 15). As if to substantialise the regulation theoretical hypothesis of a closer link between the production model and consumption practices of the postwar decades, Bourdieu demonstrates that, from the 1960s on, French state housing policy became more oriented towards home ownership, that the resulting reduction in the supply of accessible rented property redirected a section of potential tenants towards ownership and that, among home owners, social differences according to economic and cultural capital were reproduced in the process.

The trend towards living in suburbia was associated with the new housing standards. Defining cities as blighted and rural areas as backward, the New Deal programmes intended to produce a society of almost inconceivable homogeneity by modifying the physical landscape through the creation of new environments 'fit for life and the living'. These environments would merge 'the best of the urban and the rural; ideally, they would yield a pristine, homogenized, nationwide, quasi-suburban existence [...] in which America and Americans would achieve free and full expression' (Marcus and Segal, 1989, p. 264). Suburbanisation led to increasing demands for geographic mobility, and this, in turn, began to undermine local, municipal and neighbourhood relationships. Workers' housing estates and working class' areas, with the once typical atmosphere of 'belonging together' and practical solidarity based upon the principle of mutual support – which the Prussian oligarchy, for example, had monitored and suppressed – were displaced and, step by step, substituted by dormitory towns. Suburbanisation was thus tantamount to an extension of the Taylorist time–space matrix to the individual or private sphere: as more and more people ceased to live where they worked, the inner cities became too deserted immediately after shop-closing time. Alongside electric power, which made the benefits of city life available in suburban areas, it was the automobile and individual mobility that enabled the decentralisation of the population intended in New Deal programmes. While in pre-Fordist periods suburbanisation had been constrained by the need for mass transit and had congregated along streetcar and rail lines, the spreading of automobiles 'enabled urbanites to flee inner cities and settle wherever there were roads'. As a consequence, 'suburban desirability translated into an unprecedented commitment for building roadways for automobiles and for new public expenditures to make cities suitable for cars' (p. 269) – from new traffic arrangements and equipment such as one-way streets, automatic traffic signals and garages, to suburban spin-offs such as freeways and parkways. One of the main casualties of this development were public and collective transport systems. Mass-transit vehicles that generated economies of scale by carrying numerous passengers on regular routes at scheduled times were confronted with lower passenger numbers in the wake of automobiles. This transition was actively assisted by leading automobile companies such as General Motors (GM) in their competition against collective means of transport. Marcus and Siegel (ibid.) report from a 1974 hearing of the Antitrust Committee of the US Senate, according to which GM had 'helped destroy 100 street-railway systems in forty-five American cities'. Moreover, GM, Standard

Oil of California, and Firestone Tire and Rubber – three corporations that directly benefited from the increased number of motor vehicles – had formed a holding company named National City Lines, which bought street-railway companies in 16 federal states of the US. These companies were then converted to operating small, more flexible GM buses, which played to the suburban trend. The companies were subsequently resold to operators who promised to purchase only GM equipment. Similarly, from the 1920s, railways, which had operated on high-volume, long-distance and point-to-point runs, faced increasing competition and threats to their position as the pre-eminent long-haul freight carriers in the US. Gradually trains were supplanted by lorries and cars, which were deemed more in line with the new way of life.

In summary, the expansion of the Fordist mode of consumption took place against the background of the decomposition of traditional social milieux and lifestyles and was characterised by suburbanisation, increased individual geographic and social mobility and an increase in state regulation in areas such as welfare, social protection and education. In combination with the reduction of working hours, made possible by the enormous increase in labour productivity and the increasing strength of organised labour, this led, in Ulrich Beck's terminology, to a 'surge in individualisation' (*Individualisierungsschub*) away from traditional collective networks (Beck, 1992). This surge, however, also featured elements of social isolation and societal disintegration, which were accompanied by the genesis of what Hirsch and Roth (1986, p. 58) identified as 'precarious subjectivity': social homogenisation and individualisation, in combination with the emergence of forms of everyday culture such as television, which brought together masses of people who had previously led different, class- and group-specific lifestyles, crucially contributing to the relativisation of the subjective perception of class and to overlaying class awareness with shared consumption patterns. Even though quantitative levels of consumption remained stratified, the Fordist consumption norm was nevertheless generalised to the entire society, as a comprehensive normative orientation. Hirsch and Roth conclude from this that both the mental structures and the social forms of interaction came to be moulded by the consumption of commodities, which were themselves produced in capitalist ways. The promise of a constantly growing amount of use values compensated for the deformations and humiliations suffered by the individual in the Taylorist work processes and in bureaucratically regulated other areas of life. Self-perceptions of individual fulfilment and identity became increasingly dependent on the possibilities and the degree of participation in the changing standards of mass consumption (p. 59). A person's position in the work process and in the wage-labour nexus was now more than a material imperative for survival – it was the *raison d'être* of social existence as such. Pre-capitalist collective orientations were finally removed, and the 'capitalist spirit' encompassed the working class. Yet in contrast to the Calvinist period – when work itself was seen as a fulfilment and perceived as sign of being part of the 'elect', and consumption was reduced to the absolute minimum (Weber, 1986) – work was now largely perceived as an alienating activity that had to be compensated for by boundless consumption.

Hirsch and Roth (1986, p. 62) identify tensions inherent in the Fordist consumption norm by indicating that the compensation effect of the consumption of commodities can never be totally achieved. The two authors argue that 'freedom' and 'adventure' are not really experienced by smoking a cigarette: individual needs and desires that remain unmet in reality are imperfectly addressed in advertising strategies. Far removed from personal 'emancipation', the continuous consumption of ever more commodities asserted itself in the Fordist period and was sometimes perceived as an even greater heteronomy, with the corresponding loss in individual autonomy and self-determination. At the same time, it became obvious to some that mass consumption systematically produced negative external effects – for example, in the form of traffic jams, environmental problems and disasters and the desolation of cities – which began to compromise the use values of beloved consumption goods such as the car. Increasing numbers of members of the population began to question the imperatives of Fordist growth and to form ecological movements, and later green parties, which brought together many of the fragmented political groups that had come out of the student revolts in the late 1960s (Hirsch and Roth, 1986). During the same time period, new women's movements broke with the patriarchal division of labour and thereby weakened one of the cornerstones of the Fordist mode of regulation and societalisation: the nuclear family and the male-breadwinner model. This, in turn, led to changes in demands for state support, particularly from single-parent families and elderly people. Finally, these changes were accompanied by the readjustment of the spatial arrangements of Fordism: issues of housing, for example, became relevant to the extent that single households were concentrated in urban areas, which had been abandoned by the middle classes in many cases. The trend towards suburbanisation was partially reversed. However, disruptive experiences with the Fordist mode of societalisation and consumption caused only a minority to question their lifestyles. Most people simply increased their work efforts in order to achieve even higher incomes, to acquire still more commodities and to improve their chances in the competition of positional – that is, not arbitrarily augmentable – goods. The spiral of consumption and production continued to accelerate.

7
A Fossil Energy Regime

The material and energy aspect of the Fordist production and consumption model has not hitherto been in the focus of research. However, in the 1970s, Aglietta (1987, p. 118) had already proposed that a necessary condition for the success of the Fordist growth strategy was a

> revolution in energy which generalized the industrial use of energy and made possible the construction of high capacity motors which enormously increased the power available in industry. In this way, the labour process could be converted from a dense network of relationships between jobs, with intermediate products passing back and forth, and trial and error in the case of assembly, into a straightforward linear flow of the material under transformation.

During the founding stages of Fordism in the US and its later geographic expansion, the energy base of the economy was modified from biotic to fossil energy carriers. Thus, Fordism not only signified a new mode of accumulation and societalisation, but also heralded a new 'handling' of nature by human beings, whereby nature was first and foremost regarded as a source for human economic and cultural activity. The links between an industrialised economy and a corresponding energy regime developed. As Gavin Wright (1990, p. 661) observes, the continuity of the production process, closely associated to 'scientific' forms of corporate organisation, was characterised 'by "high throughput" of fuel and raw materials relative to labor and production facilities'. The volume of throughput increased with the development of the technological innovations implemented in the production process:

> sustained development of multipurpose machine tools, improvement of metals in cutting tools to increase the speed at which machines

worked, and increasing application of power to move materials more swiftly from one stage of production to the next. All three intensified the use of energy and increased the amount of capital required in the processes of production.

(Chandler, 1977, p. 279)

Taylor, for example, received patents for alloying steels and other metals. He also improved the belting that transmitted power to the machines and carried materials to the machines and their operators. The accelerated speeds, made possible by the new metals and new means of power transmission, made work processes based on older techniques of metalworking obsolete. This, in turn, enabled the introduction of further organisational changes designed to standardise, simplify and integrate work processes. So, when Henry Ford and his associates produced the low-priced Model T and then created a worldwide sales organisation, 'the resulting almost insatiable demand created a constant pressure to increase output by accelerating throughput' (p. 280).

Fossil resources in the US

The superiority of Fordism over competing growth strategies was not only due to the greater productivity growth based on mass production, but also to the easy availability and the simple and inexpensive accessibility of fossil resources in the US. The availability of cheap and reliable sources of energy and heat was required in order to substitute biotic sources of energy such as horsepower with fossil energy sources such as steam and, later, electricity, without which assembly line production would not have been possible. During the nineteenth and early twentieth centuries, all mineral raw materials required for industrialisation were abundantly available in the US: coal, ore, copper, zinc, gold, silver – and especially enormous amounts of crude oil. Not only was the US the number one producer of several minerals, but also, perhaps more importantly, the range of available resources was 'far wider than that in any other country' (Wright, 1990, p. 661). In addition, there were vast areas of rich soil, different climatic zones that enabled the industrial planting of hemp, cotton and sugar, and enormous woodlands from which huge amounts of construction timber could be extracted. In 1913, the US provided 95 per cent of the world's natural gas and 65 per cent of the world's petroleum. And, also in relation to 'copper, coal, zinc, iron ore, lead, and other minerals, [...], the United States was the world's leading producer by a wide margin' (ibid.). At the same time, the extension of a network of both railways and highways facilitated the opening up of remote areas for their economic use, the transportation of raw materials and the circulation of capital. The domestic market for both raw materials and commodities quickly grew.

Angus Maddison (1982, p. 212) assumes that the evolving US leadership in productivity was largely due to the use of advanced technology. Interestingly, he pictures Australia, which was actually the world leader in GDP per person/hour prior to the First World War, as an exception 'due largely to natural resource advantages' (Maddison, 1982, p. 258). However, Wright (1990, p. 653) points out that the US case was not different in principle from the Australian experience. Not only did post-Civil War America display 'declining relative costs of materials', but also major 'new metals discoveries continued until World War I, while the rate of discovery of new oil fields accelerated after 1900'. Wright further observes that the 'timing of leadership in industrial production coincides remarkably with American world leadership in coal production'. At the time, the US was also the world's leading producer of copper, petroleum, iron core, zinc, phosphate, molybdenum, lead, tungsten and many other materials. Not only did the US exports have a far higher natural resource content than the US imports during the period 1879 to 1940, but also this trend was 'growing both absolutely and relatively over precisely the historical period when the country was moving into a position of world industrial preeminence'(p. 658). Wright also confirms an earlier study by Lipsey (1963, p. 59), which indicated that the 'composition of manufacturing exports had been changing ceaselessly since 1879 in a fairly consistent direction – away from products of animal or vegetable origin and toward those of mineral origin'.

Limits to Fordist growth based on domestic fossil resources became manifest during the first half of the twentieth century, when the US proportion in the global production of primary energy decreased. Between 1920 and the 1950s, the US moved steadily and increasingly into the position of a net mineral importer. Not only did the extraction of raw materials in other countries expand, but several reservoirs within the US also became exhausted. The need to import raw materials increased as a result, and these extended to non-ferrous metals, bauxite, lead, zinc, copper, iron ore and petroleum. This is confirmed by Vanek's (1963) study of the natural resource content of US foreign trade between 1870 and 1955, which indicates that the US had moved from a net export to a net import position in natural resources. The US did not suddenly become 'resource poor', but the unification of world commodity markets (through transport cost reductions and the elimination of trade barriers) that accompanied the geographic expansion of Fordism made a crucial contribution to cutting the link between resources and industries. Just as American business, with the institutional help of several US administrations, was a crucial factor in this expansion, it was also 'in the forefront of the globalization of the mineral economy. In essence, the process by which the United States became a unified "economy" in the nineteenth century has been extended to the world as a whole' (Wright, 1990, p. 665). American Fordism resulted in a globalised economy of raw materials and energy (Altvater, 1992, p. 77).

The US has been a net importer of mineral resources since the postwar period; thus the reproduction of the American way of life has become more and more dependent on the extraction of mineral resources in other countries and continents. Yet, while increasing regions of the globe became involved in the provision of the energy basis of American, and then Atlantic Fordism, this led neither to socio-economic growth everywhere nor to the global spread of the Fordist mode of societalisation. One reason was that the conversion of primary energy into marketable use values largely remained concentrated in already industrialised countries. The result was that many developing countries never moved beyond the status of exporters of fossil minerals since they lacked the capacities for transforming mineral and energetic raw materials into competitive commodities. While the industrialised countries helped themselves to globally available raw materials – a fact that made a crucial contribution to increasing welfare in Western democracies and to their stabilisation – the countries whose economies focused largely on extracting raw materials remained behind. These 'extraction societies' (Altvater, 1992) became economically structured according to the needs of the industrialised countries, and this, in turn, complicated the development of economic, political and social structures that could have corresponded to the Fordist accumulation regime.

The global division of labour and the world's energy supply

This global division of labour was temporarily questioned by national liberalisation and emancipation struggles in many countries in Asia, Africa and South America, from the 1950s to the 1970s. Through strategies of industrial import substitution (Chapter 6), several countries in the Global South managed to improve their position with regard to the global distribution of wealth. Through the creation of the Organisation of the Petroleum Exporting Countries (OPEC) in 1965,[1] the oil-exporting countries demonstrated their discontent with their role as providers of cheap crude oil to maintain Fordism's fossil energy regime in the Atlantic space. On the contrary, one of OPEC's principal goals was to pursue ways and means of ensuring the stabilisation of oil prices in international markets and to contain and possibly eliminate harmful fluctuations. The price setting of oil would, in turn, lead to a stabilisation of the real incomes of the OPEC countries. The increased confidence of the OPEC countries became obvious in October 1973, when an oil embargo was proclaimed in response to the US's support of Israel's Yom Kippur War. Of similar importance was a second oil crisis, in the wake of the Iranian revolution in 1979. Even though the new regime under Ayatollah Khomeini resumed the oil exports, it remained inconsistent for a long period, forcing prices to go up. The situation worsened with the onset of the Iraqi ivasion of Iran in 1980, which led to an almost complete stop in Iranian oil production, and Iraq's production was greatly reduced. These events came to be known in the West as 'oil shocks', because they resulted in increased energy prices. This, in turn, compelled Western businesses to slow down in investment and to raise commodity prices, thereby setting an inflationary spiral in motion. The interrupted supply of cheap oil to Atlantic economies became one of the foremost crisis factors of the Fordist growth strategy; the result was a period that soon came to be associated with the label 'stagflation', since both inflation and unemployment steadily increased.

   Table 7.1 confirms that both energy consumption and throughput rose with the implementation and generalisation of the Fordist production and consumption model in the Western world. The simultaneous development of a division of labour into energy consumption countries and 'extraction' countries is equally reflected empirically in data provided by the International Energy Agency (IEA). By 1973, when the Fordist growth reached its peak, the world's total primary energy supply had risen to 6,115 million tonnes of oil equivalent (Mtoe[2]). Organisation of Economic Cooperation and Development (OECD) member states consumed more than 60 per cent of this total amount (Table 7.1). The former USSR and its allies, whose growth strategy aimed at 'surpassing' Western growth rates largely by copying Fordist production methods, were responsible for another 15 per cent of energy consumption. Although the rest of the world made up the vast majority of the world's population (see Table 7.4), it consumed only one quarter of the world's energy.

   The fact that the Fordist growth model was based on a fossil energy regime is well supported by the data (Table 7.2). Fossil fuels such as

Table 7.1 The world's total primary energy supply 1973: Regional shares (total supply: 6,115 Mtoe)

Regions	OECD	Africa	Latin America	Asia (excluding China)	China	Non-OECD Europe	Former USSR	Middle East	World Marine Bunkers
Shares (%)	61.2	3.4	3.7	5.7	7.0	1.6	14.3	1.1	2.0

Source: IEA (2008, p. 8)

Table 7.2   World's total primary energy supply by fuel shares 1973 (total supply: 6,115 Mtoe)

Fuel	Coal/peat	Oil	Gas	Nuclear	Hydro	Combustible renewables & waste	Other*
Shares	24.5	46.1	16.0	0.9	1.8	10.6	0.1

Source: IEA (2008, p. 6)
*Includes geothermal, solar, wind, heat, etc.

coal/peat, gas and oil were the sources of 86.6 per cent of the world's energy consumption in 1973, while the combined use of energy won from renewable, hydro and nuclear sources made up just 13.4 per cent. Within the OECD the dependency on fossil fuel was even greater than worldwide. Of the 3,747 Mtoe (million tonnes of oil equivalent) of primary energy (or 61.2 per cent of the world supply, Table 7.1) that the OECD countries consumed in total in 1973, 22.5 per cent stemmed from coal and peat, 18.8 per cent from gas and 52.8 per cent from oil, while the combined percentage of nuclear power, renewable, hydro and other sources made up 5.9 per cent (IEA, 2008). The general picture of a fossil energy and production regime is reinforced by a comparison of developed countries in relation to the contribution of renewables to the total energy supply in 1971 (Table 7.3). This percentage rose above 20 per cent only in the Nordic countries, while in most other countries it was well below 10 per cent. The average of the OECD countries stood at 4.8 per cent.

Carbon emissions during the period of Fordist growth

Given the limited nature of global resources of fossil fuels, doubts as to the sustainability of Atlantic Fordism and especially as to its transferability to the entire globe were formulated at an early stage. The perhaps most famous critique was launched by the Club of Rome in its attempt to delineate the *Limits to Growth* (Meadows et al., 1972). This book argued that, despite the endless propagandistic promotion of the Western way of life, its simple transfer to the rest of the world was impossible due to the limited nature of deposits of fossil resources, especially crude oil. While this critique has, with hindsight, lost little of its accuracy, it is now necessary to also raise the issue of global warming. The understanding of its origin and development remains fragmentary, if

Table 7.3 Contribution of renewables to energy supply (percentage of total primary energy supply) 1971

Country	1971
Czech Republic	0.2
Denmark	1.7
Finland	27.2
France	8.5
Germany	1.2
Hungary	2.9
Ireland	0.6
Norway	40.2
Poland	1.6
Portugal	18.9
Slovak Republic	2.4
Spain	6.4
Sweden	20.4
UK	0.1
US	3.7
OECD	4.8

Source: OECD, Economic, Environmental and Social Statistics

these are not embedded in an analysis of the upswing and generalisation of the Fordist production and consumption norm in the Western world and of the simultaneous establishment of an international division of labour in industrialised and extraction societies.

Table 7.4 provides information on the absolute numbers of the world population and CO_2 emissions and the regional percentages of world population and CO_2 emissions in 1973. The world's population was 3,917 million people in 1973, and CO_2 emissions amounted to 15,046 million tonnes. The combined percentage of the OECD, of the former USSR and of Eastern Europe in the world's population was 27.5 per cent. Yet this one quarter of the world's population was responsible for 81.9 per cent of CO_2 emissions. The Fordist production and consumption model was hence not sustainable not only for the reasons outlined in the *Limits to Growth*, but also due to the limited carrying capacity of the global ecosystems as CO_2 sinks. Climate change and its attributable disasters would have been much more advanced if developing countries in Africa, South America and Asia had not had a significantly lower share of global CO_2 emissions in relation to their proportion of the world's population. The reproduction of Atlantic

Table 7.4   World population and CO_2 emissions by region 1973

Region	World population (million)	Percentage of world population	CO_2 emissions[a] (Mt of CO_2)	Percentage of CO_2 emissions[b]
World	3917	100	15046	96.2
OECD	719	18.3	10291	65.8
Africa	390	10.0	297	1.9
Latin America	308	7.9	422	2.7
Asia (excluding China)	1258	32.1	626	4.0
China	882	22.5	892	5.7
Former USSR	250	6.4	2252	14.4
Eastern Europe	110	2.8	266	1.7

Sources: Maddison (2007, pp. 377–8) and IEA (2009, p. 45)
[a]CO_2 emissions from fuel combustion only. Excludes international aviation and international marine bunkers, which made up 594 Mt of CO_2.
[b]Excludes international aviation and international marine bunkers, which made up 3.8 per cent of CO_2 emissions.

Fordism and of Western welfare was dependent upon the fact that the vast majority of the world's population lived under economically less profitable but ecologically more sustainable circumstances. Table 7.5 presents data provided by the World Resources Institute (WRI) in relation to the CO_2 emitted per person and country.[3] Looking at the CO_2 emissions per capita of all countries combined, there was a tremendous rise from 2.32 to 4.08 tonnes in the period 1950 to 1973, the peak of the Fordist accumulation and consumption period. Fast economic growth and the increase in the combustion of fossil fuel energies led to per capita emissions of CO_2 doubling to 12.19 tonnes in the developed countries between 1950 and 1973. By comparison, individual CO_2 emissions also rose in the developing world, but to just 0.88 tonnes. Thus, in 1973, an inhabitant of the developing countries emitted little more than 7 per cent of the amount of CO_2 emitted by an inhabitant of the developed world.

The differences are also remarkable within the developed world. In the US, Fordism's homeland, per capita emissions stood at 21.84 in 1973. This was over 20 times more than the emissions of a person in an average developing country and nearly twice as much as the emissions

Table 7.5 The world's CO_2 emissions per capita 1950–1973: Developing and developed countries

	1950	1955	1960	1965	1970	1973
All countries	2.32	2.67	3.05	3.34	3.89	4.08
Developing countries	0.24	0.37	0.72	0.67	0.86	0.88
Developed countries	6.4	7.34	7.99	9.39	11.22	12.19
Selected Developed Countries						
Australia	6.74	7.79	9.18	10.26	12.41	13.30
Canada	11.32	10.90	10.62	13.16	15.57	16.53
Denmark	5.25	6.24	6.87	9.38	12.63	11.73
Finland	1.73	2.70	3.51	5.68	9.00	10.73
West-Germany	7.56	10.44	11.38	12.90	13.00	13.68
Hungary	2.01	3.69	4.62	6.27	5.53	6.29
Ireland	3.35	4.13	4.87	5.33	6.96	7.55
Netherlands	5.10	5.72	6.12	7.83	10.05	11.21
Norway	2.67	3.45	3.87	4.68	6.71	6.44
Poland	4.57	5.72	6.94	7.90	9.41	10.19
Portugal	0.70	0.80	0.79	1.07	1.68	2.05
USSR (Russian Federation)	4.22	5.93	7.77	9.65	1.39	12.62
Spain	1.22	1.45	1.80	2.34	3.48	4.46
Sweden	4.23	5.56	6.62	8.30	11.83	11.10
UK	9.98	11.39	10.95	11.42	11.70	11.89
US	16.21	16.09	15.42	17.13	20.36	21.84

Note: In tonnes of CO_2e
Source: WRI

of an average person in other developed countries. Generally, the quantity of individual CO_2 emissions follows the economic development level of a country. Spain and Portugal, which display the lowest CO_2 emissions per person among the countries listed as 'developed' by the WRI, were barely industrialised in 1973. Both countries were ruled by dictatorships that aimed at economic autarky and cut their countries off from Western culture and consumption styles as much as possible. Similarly, the Republic of Ireland has never been fully industrialised and remained at the economic periphery of Europe until the 1990s. On the other hand, countries whose individual CO_2 emissions were above the average of the developed countries include the Federal Republic of Germany, Canada, Australia and the former USSR. The emulation of the Fordist accumulation model by the 'real-existing' socialist world is reflected in individual CO_2 emissions, which, in the case of the former USSR, Poland and Hungary, were on similar levels to those in Finland, the Netherlands and the UK. It was not only an economic, but also an ideational and hegemonic weakness that – despite existing expertise from oppositional economists, who were normally marginalised and/or persecuted – the Eastern bloc never managed to develop an autonomous socialist growth strategy that could have been more than just an undemocratic and bureaucratic copy of Western Fordism and could have given ecological parameters a higher priority.

Part III
Finance-Driven Capitalism

There is great agreement among authors in the fields of political economy, industrial relations and social policy that the Fordist accumulation regime's potential for profitability was exhausted in the course
of the 1970s (Koch, 2006a, pp. 30–4). But there is still no consensus
over a precise definition of a 'post-Fordist' regime. During the 1990s,
many authors emphasised the various ways of organising flexibility
and the different methods and scope of labour market coordination
when understanding recent growth strategies from a comparative perspective. Over the last decade, a growing number of scholars have
focused on the idea of the 'financialisation' of socio-economic relations in order to understand the current period of capitalist growth
(Boyer, 2000; Huffschmid, 2009; Krugman, 2009; Stockhammer, 2008).
This concept covers a range of phenomena, including the deregulation
of the financial sector and the liberalisation of international capital
flows, significant increases in financial transactions and the prolifera-
tion and profitability of new financial instruments and investors such
as hedge funds. It is hypothesised that priorities in companies' competitive strategies moved away from investments in the real economy
towards greater focus on financial profits, financial markets and foreign investment. At the same time, the increased significance of the
financial sector amplifies the vulnerability of economy and society to
the effects of severe financial crises. I agree with the above-mentioned
authors that the most recent period of capitalist growth is best understood in terms of a 'finance-driven' accumulation regime; I will delineate
its rise in historical and empirical terms (Chapter 8). Chapter 9 follows the development of the international division of labour under the
new circumstances of market liberalisation and financial globalisation.
Chapter 10 analyses the development, modification and expansion of
the consumption norm that emerged under Fordism. Chapter 11 deals
with the energy regime of the new growth strategy and raises the issue
of whether its fossil-dependent nature has changed qualitatively. As in
Chapter 7, special emphasis will be placed on carbon emissions and
climate change.

8
The Rise of a Finance-Driven Accumulation Regime

During the course of the 1970s the Fordist gross domestic product (GDP) and productivity growth model weakened. Aglietta (1987, p. 119) describes this as a consequence of the Taylorist logic itself: 'the further the fragmentation of individual tasks and integration of jobs by the machine system have already been taken, the more costly in means of production is subsequent intensification of the output norm'. The reason was the great technical rigidity of the mechanised system. Further increases in labour productivity were contingent upon additional investments in fixed capital on an ever greater scale. At the same time, the associated search for increases in productivity via work intensification led to growing alienation on the shop floor. Workers' protests began to undermine companies' profitability from the late 1960s on, when the almost total capacity utilisation of fixed capital concentrated great numbers of workers on the same shop floor, carrying out very similar tasks. Hence a numerically growing and relatively homogenous working class emerged as a result of the Taylorist organisation of the work process, which then became the most effective obstacle to Fordist growth (Jessop, 2002, pp. 81–2). This is reflected in Tables 8.1 and 8.2. Labour productivity growth fell in all countries of the Atlantic space, when compared to the levels of the 1960s. This fall was especially steep in most of the European Union (EU) 15 countries, where a 4.8 per cent average growth in labour productivity in the 1960s slowed down to a mere 0.8 percentage growth in the 2000s. In countries such as Japan, which had especially benefited from the postwar boom, the deceleration of labour productivity growth was especially great. In contrast, in the UK, where Fordist growth had been flawed in the first place, this slowdown was less evident.

Table 8.1   Labour productivity growth in selected European countries, US and Japan: 1961–2010

	1961–1970	1971–1980	1981–1990	1991–2000	2001–2010
Austria	5.1	2.9	2.0	2.0	1.0
Belgium	4.4	3.2	1.8	1.5	0.6
Denmark	3.5	1.9	1.6	2.2	0.6
Finland	4.4	3.5	2.5	2.8	1.2
France	5.3	3.2	2.1	1.6	0.8
Germany*	4.2	2.6	1.3	2.5	0.9
Italy	6.2	2.8	1.8	1.6	−0.1
Netherlands	3.9	2.5	1.6	1.5	0.9
Sweden	3.9	1.2	1.5	2.6	1.4
UK	2.5	1.7	2.0	2.3	1.0
Ireland	4.2	3.7	3.8	3.3	1.6
Portugal	5.6	4.9	· 3.5	2.2	0.5
Spain	6.7	4.2	1.9	1.1	1.0
EU 15	4.8	2.8	1.8	2.0	0.8
US	2.3	1.2	1.4	1.8	1.7
Japan	8.6	3.7	3.7	1.0	1.1

Note: Gross domestic product at 2000 market prices per person employed; annual percentage change.
*1961–91: West Germany.
Source: EU Commission: *European Economy*, Autumn 2009: Statistical Annex

On the demand side, the only barrier to Fordist growth appeared to be the existing scale of production. This was due to the all but inexhaustible demand for durable consumer goods after the destruction of European economies during the Second World War. However, geographically, markets were largely limited to the national scale, where only so many cars, refrigerators and washing machines could be sold. In time it became obvious that domestic markets were not 'growing' sufficiently fast to accommodate expanding capital valorisation and the simultaneously growing productivity of work, but remained at best stable (Revelli, 1997). Once the demand for mass-produced commodities could no longer be relied upon, sales and the market ceased to be secondary and dependent variables and became primary and independent. Management strategies changed accordingly. In an attempt to achieve further economies of scale and to compensate for the relative market saturation of their home markets, companies began to expand into foreign markets. This process was facilitated by the emergence and expansion of new information and communication technologies such as personal computers and the Internet, which enabled capital to become increasingly

Table 8.2 GDP growth in selected European countries, US and Japan: 1961–2010

	1961–1970	1971–1980	1981–1990	1991–2000	2001–2010
Austria	4.7	3.6	2.2	2.5	1.4
Belgium	4.9	3.4	2.0	2.2	1.2
Denmark	4.4	2.9	2.0	2.2	1.2
Finland	4.8	3.8	3.0	2.0	1.6
France	5.7	3.7	2.4	2.0	1.2
Germany*	4.4	2.9	2.3	2.1	0.6
Italy	5.7	3.8	2.4	1.6	1.2
Netherlands	5.1	3.0	2.3	3.2	1.1
Sweden	4.6	2.0	2.2	2.0	1.4
UK	2.8	2.0	2.7	2.5	0.9
Ireland	4.2	4.7	3.6	7.1	2.6
Portugal	5.8	4.9	3.6	3.0	2.8
Spain	7.3	3.5	2.9	2.8	2.0
EU 15	4.8	3.1	2.4	2.2	1.1
US	4.2	3.2	3.2	3.4	1.7
Japan	10.1	4.4	4.6	1.2	0.5

Note: Gross domestic product at 2000 market prices; annual percentage change.
*1961–91: West Germany.
Source: EU Commission: *European Economy*, Autumn 2009: Statistical Annex

mobile. It became also possible to reorganise the production process on a transnational scale and to utilise foreign credit and tax havens to reduce the costs of both borrowing and transfer payments (Fröbel et al., 1981). The transnationalisation of production and circulation networks had, in turn, major implications for the financial base underpinning Keynesian economic and social policies, since national governments could no longer act as if their own economic space was hermetically enclosed. Increasingly, governments had to attract transnational capital by offering favourable tax and social security conditions, thereby competing with other locations. The fiscal crisis of the state was aggravated by the fact that more and more nationally and domestically focused companies were paying less tax, since they were generating diminishing gross profits as a result of the increased bargaining power of national trade unions. The combined effect of supply and demand factors resulted in a long-term decline of GDP growth in the Atlantic space. As indicated in Table 8.2, growth rates of the GDP fell in all major capitalist countries when compared to the levels of the 1960s. In the case of the EU 15 countries, this decline has been continuous, going from 4.8 per cent to 1.1 per cent in the 2000s. Although Japan (in the 1980s) and the US (in the

1990s) enjoyed short-lived revivals in GDP growth, their performances in the 2000s were also well below that of the Fordist period. In Japan, the decrease in annual GDP growth was especially drastic (from over 10 per cent in the 1960s to a mere 0.5 per cent in the 2000s).

The growth crisis of Fordism led to profound changes in the international regulation of the world economy. The US productivity advantage eroded as European and Japanese exports increased at a faster rate than those of the US. At the same time, US direct investments expanded overseas, which resulted in a further outflow of the dollar from the US, mainly towards Europe. In the US, a rapidly growing deficit in the balance of payments emerged and undermined the commitment to exchange foreign-held dollars for gold. As the deficit in the balance of payments increased, the gold standard lost its relevance. By the early 1970s, a devaluation of the US dollar seemed inevitable. When the US government withdrew its commitment to convert foreign dollars into gold, the international currency system negotiated at Bretton Woods effectively ended (Bordo and Eichengreen, 1993). The system of exchange rates now became more flexible. The use and the exchange rate of the US dollar in the currency markets declined. The dollar did not lose its function as a world currency but was supplemented by the German mark and the Japanese yen, and, more recently, by the euro; these have taken on the role of reserve currencies. With regard to international trade, the transition to flexible exchange rates resulted in greater market uncertainty, since currencies were now free-floating. At the same time, the potential for currency speculation grew enormously and new investment possibilities emerged for financial capital, becoming increasingly more available as the crisis of Fordism unfolded. Further opportunities for capital valorisation opened up as the new information and communication technologies facilitated the ready availability of necessary information on markets and permitted the purchase of blocks of shares anywhere in the world. Financial transactions could now be undertaken 24 hours a day in different time zones.

Neoliberalism and re-regulation

The transition from a Keynesian to a neoclassical economic paradigm was not in the first instance a matter of the 'best argument' within economic theory; it reflected instead a political and ideological shift in power relations in the wider society. Not so long ago – 40 years approximately – what came to be known as neoliberalism was just a minority voice within the discipline of economics. David Harvey (2009, pp. 19–23) recalls the story of the Mont Pelerin Society (named after the Swiss spa where the group originally met), which assembled an exclusive group of then heterodox political philosophers (among them Friedrich von Hayek and Karl Popper) and economists (including Ludwig van Mises and Milton Friedman) in the late 1940s. The group regarded itself as being 'liberal' on account of its commitment to ideals of personal freedom. However, it also signalled its 'adherence to those free market principles of neo-classical economics that had emerged in the second half of the nineteenth century' (Harvey, 2009, p. 20), especially in the case of Alfred Marshall and Léon Walras. Neoliberal doctrine was understood as following in the footsteps of Adam Smith and of the idea that the 'invisible hand' of the market was the best way of mobilising individual motivation and interest. This would be to the benefit of all, over the medium and long term. This commitment of the group was in opposition to all kinds of state 'interventionist' economic theories, and especially to John Maynard Keynes and his followers, who, after the Great Depression, tried to keep the business cycle and recessions within certain limits through the active role of the state in socio-economic regulation. State decisions, argued the group, 'were bound to be politically biased depending on the strength of the interest group involved (such as unions, environmentalists, or trade lobbies)'. They were also bound to be wrong in 'matters of investment and capital accumulation [...] because the information available to the state could not rival that contained by market signals' (Harvey, 2009, p. 21).

Neoliberalism has never been a particularly coherent theoretical framework: the

> scientific rigour of its neoclassical economics does not sit easily with its political commitment to ideals of individual freedom, nor does its supposed distrust of state power fit with the need for a strong and if necessary coercive state that will defend the rights of private property, individual liberties, and entrepreneurial freedoms.
>
> (Harvey, 2009, p. 21)

Nor was it of practical relevance prior to the mid-1970s. It was only in the wake of the Chilean *coup d'état* in 1973, carried out with the material and ideological assistance of the US, that neoliberal ideas became economic and political practice (Koch, 1998b, 1999). Many young Chilean economists had studied at the Chicago School of Economics, where Milton Friedman played the leading role, defiantly resisting the Keynesian mainstream.[1] It was only when Pinochet's military government removed the disruptive effect of trade unions and oppositional parties that the 'Chicago boys' were able to carry out 'supply-oriented' reforms. Chile was in fact used as a testing ground for neoliberal theories. Many of the measures subsequently exported to other parts of the world had been tried and tested in Chile: the break with industrial import substitution as a development path for developing countries; the privatisation of public firms and the deregulation of prices, capital and labour markets; and, last but not least, the creation of favourable conditions for the investment of financial capital. But it was only in the 1980s that neoliberal and monetarist approaches became the new doctrine of public policy.[2] After the election of Reagan and Thatcher in 1979, privatisation and the creation of favourable conditions for the investment of financial capital were seen as crucial to socio-economic regulation. For Thatcher's conservative government, it was obvious that Keynesianism and the maintenance of full employment – a commitment of all British postwar governments – was to be abandoned and that monetarist supply-side solutions were to be welcomed in order to cure the stagflation that characterised the British economy in the 1970s. However, paradoxically, liberalising market forces required a strong state, which first intervened in industrial relations and the wider economy in order to make itself redundant at a later stage. The most important cornerstone of the new economic and political edifice was the weakening of collective forces and in particular of the power of the unions, especially at company level (Sakowsky, 1992). At the same time, public expenditure was reduced, particularly through the privatisation of state companies and the diminishment of public employment. Reforms of the welfare system included various reductions of the replacement ratio for unemployment benefits, and a tighter link between benefit entitlements and the willingness to take up work or training (Rhodes, 2000, p. 46). These measures were underpinned with relentless ideological assaults on anything remotely evocative of 'collectivism'. Indeed, for Thatcher, the 'object' was to 'change the soul', while 'economics' was the 'method' (Margaret Thatcher in an interview in the *Sunday Times* on 1 May 1981). In the US, Paul Volcker, at the head of the Federal Reserve (FED), was equally determined to 'break the back of inflation' as well as the strength of organised labour. New Deal principles in fiscal and monetary policies and full employment were abandoned as an important goal of socio-economic regulation (Harvey, 2009, p. 23). The FED base rate rose from an average of 8 per cent in 1978 to over 19 per cent at the beginning of 1981 and did not return to less than double digits consistently until after 1984. Since the 'Volcker shock', as it became known, the FED has explicitly guaranteed the interest rate in order to keep inflation at very low levels. Panitch and Konings (2009, p. 33) point out that this attack on inflation was only effective 'in combination with the strong underlying capacities of the American economy: its technological base, depth of financial institutions, and the resources that came with its imperial role'. By breaking the inflationary spiral, the US not only won back the confidence of financial markets, it also put itself in the position of being able to tell other states to change their policy priorities too. The FED's unwavering anti-inflationary commitment became the basis of a new dollar-based world economy. The further liberalisation of the US's own financial markets not only 'deepened the domestic strength and liquidity of these markets' but also 'supported their further internalization. It was this that now crucially sustained the dollar as an international currency and made US government securities seem as good as (indeed, because they paid interest, better than) gold' (Panitch and Konings, 2009, p. 33).

In this new system, based on open capital markets and floating exchange rates, the US dollar remained the main global monetary unit of account and the main means of payment for oil and other crucial products. Consequently, swings in the value of the dollar measured in other currencies did not change prices for these products for those operating in the dollar zone. Furthermore, due to the dollar's character as 'world money', the US could 'run up current account deficits and enormous external debts without facing the kind of monetary payments constraints facing other states' (Gowan, 2004, pp. 66–7). Since the dollar was now freed from the constraints of 'any international anchor and rules common to all', the US could subordinate international monetary conditions to suit its perceived interests: 'When the United States was in recession, the U.S. authorities could drive the dollar down to generate an export-led revival; when the United States was rising into boom, the Treasury Department could swing the dollar up massively against other currencies' (p. 66). A range of mechanisms was established to make foreign markets more assessable to US companies and to transform the internal and external environments of states in order to make them more adaptable to US political and economic priorities. The most influential forces determining the rules of this regime were the US treasury department, the FED[3] and the private financial companies located on Wall Street. Peter Gowan points out that the relations between the pillars of the new power elite were characterised by personal exchange, close working ties and similar goals. The regime's institutional coherence was strengthened by the compatibility between policy priorities of international organisations – such as the International Monetary Fund (IMF), the World Bank (WB), the World Trade Organisation (WTO), the Bank of International Settlements, the Basle Committee and the International Organisation of Securities Commissions – and the economic strategies and interests of US elites. In fact, Gowan presupposes a division of labour between US banks and the IMF, in which 'the IMF covers the risks and ensures that the US banks don't lose (countries pay up through structural adjustments etc.) and flight of capital from localized crises elsewhere ends up boosting the strength of Wall Street' (Gowan, 1999, p. 35).

In the US, several federal acts facilitated the liberalisation of international capital flows and deregulated various social policy fields. In housing and mortgage financing, the annulment of the Glass–Steagall Act in 1999 was of special importance. This act had made a distinction between commercial banks, which accepted deposits, and investment banks, which did not. Commercial banks were 'sharply restricted in the risks they could take; in return, they had ready access to credit from the FED (the so-called discount window), and, probably most important to all, their deposits were insured by the taxpayer' (Krugman, 2009, p. 157). Investment banks, on the other hand, were less tightly regulated. This was seen as acceptable, since being non-depository institutions they were not thought to be subject to bank runs. Krugman (ibid., p. 160) refers to the investments banks in terms of a 'parallel' or 'shadow banking system', which 'expanded to rival or even surpass conventional banking in importance'. Simultaneously, the private sector began to play a more important role in banking supervision, both at the national and at the international level. In continental Europe, measures to facilitate transnational capital flows were taken at European level: the Maastricht Treaty, the Stability and Growth Pact and the Services Directive. These policy shifts were later made compatible with employment and social policy objectives. Far from being a synonym for deregulation, supply-side economic management involved active re-regulation at different (national, transnational, regional) levels (Jessop, 1999) and was usually implemented by a strong state (Koch, 2008). In all its national varieties, then, neoliberalism in general and the opening up of domestic markets to transnational capital in particular were outcomes of conscious political decisions. A good overview of the directions of domestic regulatory changes is provided by United Nations Conference on Trade and Development's (UNCTAD's) *Database on National Laws and Regulations*. This contains information on the number of countries that introduced changes in relation to foreign direct investment (FDI)

Table 8.3   National regulatory changes in relation to FDI: 1992–2008

	1992	1996	2000	2004	2008
Number of countries that introduced changes	43	66	70	103	55
Number of regulatory changes	77	114	150	270	110
More favourable to FDI	77	98	147	234	85
Less favourable	0	16	3	36	25

Source: UNCTAD (2009, p. 31)

in comparison to the previous year, and on whether these changes facilitated or complicated FDI (Table 8.3).

UNCTAD identifies an increase in national regulation regarding FDI from 77 cases in 43 countries in 1992 to 270 cases in 103 countries in 2004. By 2008 – the year in which the global financial crisis struck – the number of changes in national regulations had decreased to 110. These changes occurred in 55 countries. Their overwhelming majority made FDI more favourable. While in 1992 there was no regulation change that would have made FDI less favourable at all, by 2008 the number of such regulations was 25 in 110 cases. In other words, while little more than one-fifth of regulatory changes made FDI more difficult, almost four-fifths made it easier. The countries that introduced measures that complicated FDI included Algeria, Bolivia, Peru and Venezuela (UNCTAD, 2007, p. 15), where reforms strenghtened the position of the national state in the extractive industries, especially with regard to foreign private investors. In Russia and China, the governments identified so-called 'strategic sectors', within which FDI inflows were limited by regulation. On the whole, however, the UNCTAD data leave no doubt that regulatory modifications at national level overwhelmingly facilitated market liberalisations and openings for FDI and transnational corporations. At international level, developed countries provided developing countries with unpredecented market access, especially for primary and value-added products, in exchange for lower tariffs and compliance with three new trade-related agreements: the Agreement on Trade-Related Investment Measures (TRIMS), the General Agreement on Trade in Services (GATS) and the Agreement on Trade-Related Aspects of Intellectual Property Rights (TRIPS). Thanks to TRIMS, developing countries lost much of their ability to define the requirements placed upon multinational corportations, including prohibiting the use of domestic content requirements, trade balancing, foreign exchange restrictions and local procurement requirements for public agencies. GATS weakened the ability of states to protect and regulate the service sector. The agreement included both private goods such as tourism and financial services and public goods such as health care, education and water and sanitation. Countries were required to identify those areas within the service sector they intended to protect, as opposed to identifying a 'negative list' of areas in which international competition was unlimited (Roberts and Parks, 2006, p. 53). Finally, the TRIPS agreements restricted states' abilities to refuse patents for certain products, fixed the period of patent coverage at 20 years, limited third-party access to patented products and made non-compliance punishable through the WTO's dispute resolution mechanism.

Falling wage shares and the financialisation of investment

The domestic spending power of wage-earners in the Western centres of a financialised capitalist economy is less important than in Fordism. Commodities and services are increasingly offered and sold worldwide, which makes profitability less dependent on markets in the Atlantic space. At the same time, the decrease in real wages and the corresponding fall in the spending power of wage-earners are partially compensated by the increased access of the wage-earning class to consumer loans. Financialisation is, therefore, likely to be accompanied by a redistribution of primary income, from the wage-earning to the capital-owning class. Indeed, Marx (1974, p. 420) followed David Ricardo in defining the mutual relationship of the two main social classes not on the basis of their absolute incomes, but according to the relative shares of profits and wages in the annual GDP (Koch, 2010). Thus, the relative strength or weakness of the wage-earning class vis-à-vis employers does not follow from income in absolute terms, but from *proportionate* wage. This is illustrated in Table 8.4, which deals with the 'wage share' as a percentage of GDP in the EU, the US and Japan for the period 1961–2010.

The relative position of the wage-earning class vis-à-vis employers has deteriorated in Europe, US and Japan since the 1970s. In EU 15, the wage share decreased from 72.9 per cent in the 1970s to 65.6 per cent in the 2000s. It fell in every EU member state, Austria, Ireland and Portugal displaying especially large declines. In the three Eastern European countries – Poland, Czech Republic and Hungary – the level of the wage share was even below that of the EU 15 mean, with a tendency of declining even more. In Japan and the US, the wage share decreased from 76.6 and 69.9 per cent in the 1970s to 65.7 and 65.1 per

Table 8.4   Wage share in selected European countries, US and Japan: 1961–2010

	1961–1970	1971–1980	1981–1990	1991–2000	2001–2010
Austria	79.8	80.3	76.0	71.4	65.4
Belgium	63.5	70.2	70.4	70.7	69.0
Denmark	67.8	70.4	67.4	66.4	63.8
Finland	75.8	73.4	71.4	66.3	62.8
France	73.8	73.9	72.2	67.2	65.6
Germany*	67.8	70.4	67.4	66.4	63.8
Italy	72.5	72.2	68.7	64.6	62.2
Netherlands	67.8	73.8	69.2	67.5	66.1
Sweden	74.2	76.2	72.0	67.6	69.0
UK	72.9	74.3	72.8	71.9	70.8
Ireland	78.0	76.0	71.3	62.4	56.2
Portugal	74.8	84.4	67.4	70.2	71.7
Spain	70.4	72.3	68.3	66.8	62.3
EU 15	71.5	72.9	70.2	67.6	65.6
Poland	–	–	–	66.5	57.0
Czech Republic	–	–	–	55.6	58.4
Hungary	–	–	–	61.7	61.3
EU 27	–	–	–	66.5	65.3
US	70.0	69.9	68.3	67.1	65.1
Japan	72.8	76.6	73.0	70.1	65.7

Note: As percentage of GDP at current factor cost.
*1961–91: West Germany.
Source: EU Commission: *European Economy*, Autumn 2009: Statistical Annex

cent in the 2000s. The converse argument is that an increasing share of the newly added product was appropriated in the form of profits in all the main capitalist countries. Yet the rising level of profit appropriated by entrepreneurs could only be partially reinvested in productive industries, as there was not enough spending power left at the 'bottom' of the income scale to purchase the required number of products. Thus the general tendency for the finance-driven period was for profits to rise at the expense of wages, while the investment rate decreased simultaneously (Table 8.5).

In the major economies (Germany, France, the UK and US), the investment/profit ratio has displayed a declining trend since the 1970s. The investment rate increased only in Denmark, remaining constant in Spain. Stockhammer (2008) observes that Keynes would probably not have been surprised by the fact that increasing profits did not result in additional investment made by companies, even though the increased access to new financial instruments would have enabled companies to

Table 8.5   Investment as percentage of operating surplus in selected countries

	1970s	1980s	1990s	2000s
Denmark	46	47	46	49
Germany	52	48	42	43
Spain	47	40	44	47
France	46	46	42	43
Ireland	50	44	30	28
Netherlands	48	39	38	38
Finland	57	57	41	36
Sweden	59	52	46	51
UK	55	48	44	42
EU*	47	44	40	40
US	46	44	39	39
Japan	58	59	61	56

Note: Private gross fixed capital formation.
*Unweighted average of available EU countries.
Source: Stockhammer (2008, p. 190)

make additional investments. This is because he regarded investment as a result of expected profits rather than as a result of real profits. Another factor that explains diminishing investment rates is the increased role of shareholders in companies (Stockhammer, 2008, p. 191): 'rather than a management-labour balance (like in the Fordist era), firms are now characterized by a management-shareholder balance'. Given the expanding disposability of financial instruments, shareholders were more likely to invest in financial markets than in physical projects. Stockhammer (2008, p. 191) adds that, due to the increased volatility on financial markets, in particular the volatility of exchange rates, 'firms face a higher degree of uncertainty'. This led to a reduction in investment into manufacturing, while unproductive investment into investment banks, investment and equity funds, insurances and pension funds became more attractive. Some of these new instruments, such as equity funds, were first private and unlisted funds that concentrated and controlled private assets. These assets were used to buy the majority of a company's shares and to restructure its management and organisational structure so that its stock prices rose over the short term. The purchased company, thus, became an attractive object for resale. Later, new financial instruments such as hedge funds entered the scene (UNCTAD, 2007, p. 7); their number grew rapidly in the 1990s and 2000s (Stiglitz et al., 2006, pp. 6–7). Krugman (2009, p. 120) defines hedge funds as 'investment

Table 8.6 Development of the global nominal GDP and financial stocks: 1980–2006 ($trillion)

	1980	1990	1995	2000	2006
Nominal GDP	10	22	29	32	48
Financial Assets	12	43	66	94	167

Source: Farrell et al. (2008, p. 3)

institutions that are able to take temporary control of assets far in excess of their owner's wealth'. Often operating offshore, thereby liberating themselves from taxes and public charges, hedge funds often realised spectacular profit rates by going 'short' in some assets, promising 'to deliver them at a fixed price at some future date', and going 'long' in others. 'Profits come if the price of the shorted asset falls (so that they can be delivered cheaply) or the value of the purchased asset rises, or both' (ibid.). The next step was that funds not only turned their clients' assets and loans into tradable commodities but also were listed themselves, in order to attract further financial capital. 'Funds of funds', that is, funds that invest in other funds, became important players in international capital markets (Koch, 2009).

The trend towards financial investment and the corresponding accumulation of financial assets is clearly reflected in Table 8.6. Between 1980 and 2006, financial assets – the sum of the value of all bank deposits, government debt securities, corporate debt securities and equity securities – grew continuously; and they did so much more quickly than the nominal global GDP. While the value of financial assets rose from $12 trillion to $167 trillion or by factor 14, global nominal GDP only grew from $10 to $48 trillion or by factor 4.8. In 1980, the stocks of financial assets and nominal GDP were roughly the same size. By 2006, however, the value of financial stocks was three and a half times more than that of the global nominal GDP.

9
The Recomposition of the International Division of Labour

The trend towards the financialisation of investment was accompanied by its transnationalisation. Strikingly, between the 1980s and 2007, all Foreign Direct Investment (FDI) indicators selected for Table 9.1 demonstrate much greater growth rates than those of the world gross domestic product (GDP). It was only in the crisis year 2007/2008 that the world GDP continued to grow while all other indicators were reduced. More particularly, the annual growth rates of FDI inflows oscillated between 20 and 50 per cent between the 1980s and 2007. The FDI-inward stock expanded from $790 billion in 1982 to $15,660 billion in 2007, by factor 20. The sales of foreign affiliates rose by factor 12 from $2,530 billion in 1982 to $31,763 billion in 2007. During the same period, the sales, gross product and total assets of foreign affiliates increased constantly. Employment by foreign affiliates also grew by factor 4. However, the employment growth rate in transnational corporations was slower than that of the FDI stock and than that of the gross product of foreign affiliates. This indicates a transition from employment-extensive towards capital- and knowledge-intensive work processes. The increase in FDI up to 2007, according to UNCTAD (2007, p. 4), was a reflection of the increase in corporate profits and stock prices, which increased the value of cross-border Mergers and Acquisitions (M&As). The latter accounted for a large share of such flows. Between 1990 and 2007, the annual value transferred through such fusions increased by factor 9, from $112 billion to $1,031 billion. As with the other FDI indicators, M&As shrank by one-third during the financial crisis of 2008. The crisis led to a general slowdown in the transnationalisation and internationalisation process of global investment. The future development is presently unclear.

Acquisitions of companies by other companies, buyouts of foreign affiliates, trading in complex derivatives and the emergence and growth

Table 9.1 Foreign direct investment: Selected indicators 1982–2008

	Value at current prices (billion $)				Annual growth rate (per cent)						
	1982	1990	2007	2008	1986–1990	1991–1995	1996–2000	2005	2006	2007	2008
FDI inflows	58	207	1979	1697	23.6	22.1	39.4	32.4	50.1	35.4	−14.2
FDI inward stock	790	1942	15,660	14,909	15.1	8.6	16.0	4.6	23.4	26.2	−4.8
Sales of foreign affiliates	2530	6026	31,764	30,311	19.7	8.8	8.1	5.4	18.9	23.6	−4.6
Gross product of foreign affiliates	623	1477	6,295	6020	17.4	6.8	6.9	12.9	21.6	20.1	−4.4
Total assets of foreign affiliates	2036	5938	73,475	69,771	18.1	13.7	18.9	20.5	23.9	20.8	−5.0
Employment by foreign affiliates (in thousands)	19,864	24,476	80,396	77,386	5.5	5.5	9.7	8.5	11.4	25.4	−3.7
Cross-border mergers and acquisitions	–	112	1031	673	32.0	15.7	62.9	91.4	38.1	62.1	−34.7
World GDP	11,963	22,121	55,114	60,780	9.5	5.9	1.3	8.4	8.2	12.5	10.3

Source: UNCTAD (2009, p. 18)

Table 9.2 Domestic debt and GDP in the US ($trillion)

	Gross Domestic Product	Total Debt	Financial Firm	Non-Financial Business	Government (local, state, federal)	Households
1970	1.0	1.5	0.1	0.5	0.4	0.5
1980	2.7	4.5	0.6	1.5	1.1	1.4
1990	5.8	13.5	2.6	3.7	3.5	3.6
2000	9.8	26.3	8.1	6.6	4.6	7.0
2007	13.8	47.7	16.0	10.6	7.3	13.8

Source: Foster and Magdoff (2009, p. 121)

of hedge funds involved a significant amount of borrowing, thereby adding to an increasing amount of debt in the economic system. As profits in the real economy diminished and new financial instruments and markets opened up, leverage debt was bound to increase. This is indicated in Table 9.2, which compares levels of GDP and debt in the US between 1970 and 2007. In 1970, the GDP and total debt were of roughly the same order of magnitude. Total debt has increased to a greater extent than the US GDP since then. Although diverse companies, the state and households increased their borrowing, the debt explosion of financial firms over the last two decades is particularly evident. This debt increased by factor 7 from $2.6 trillion in 1990 to $16 trillion in 2007, confirming Foster and Magdoff's (2009) thesis of a qualitative shift within the banking sector: gone are the days when banks were 'primarily loaning funds that had been deposited by the public' and when banks 'collected interest and principal from those who has taken on debt and paid a share to depositors' (Foster and Magdoff, 2009, p. 121). Today's banks have themselves become massive borrowers: 'Financial institutions of all types now accumulate huge quantities of debt as they attempt to make money with borrowed money'. Recalling the falling investment rates in Western countries (Table 8.5), one must indeed come to the conclusion that debt undertaken by financial institutions has ceased to have the main purpose of stimulating production. Its primary purpose has become the continuous generation and accumulation of speculative profits.

The growing debt of US companies and government authorities was the result of yet another trend. Many authors have alluded to the fact that, from about 1980, the US became the 'consumer of last resort' in the global economy. This means that US citizens consume generally more than they produce. While in some industries the US maintained its role as world market-leader in terms of technological development and product quality, foreign imports proved superior to US products in many other industries. The resulting huge foreign trade deficit functioned like a motor for booming export industries like China and for some of the EU core countries, especially Germany.[1] For the US, however, this meant growing levels of debt in relation to its trading partners. The situation was aggravated by the fact that the American saving rate tended towards zero, so that the country became dependent on capital inflows from foreign investors who bought bonds from companies or public institutions, shares or other stocks. The resulting income was used to serve the US's deficits and foreign debt. The country had to absorb a significant share of global savings in order to make up for its current account deficit (Bischoff, 2008). Like Britain, which reached the height of its financial power after its decline as 'workshop of the world', the US's financial power outlived the undisputed industrial pre-eminence that had characterised Fordism. In contrast to Britain, however, 'the US benefited rather than suffered from its transformation from an international creditor to an international debtor' (Panitch and Konings, 2009, p. 3).

The deregulation of finances and trade and the geographic and technological reorganisation of the production process had different regional consequences across the globe. Those countries that had obtained loans at favourable interest rates in the early 1970s, particularly in Eastern Europe and South America, experienced the disastrous consequences of the drastic rise of interest rates in the US following the 'Volcker Shock'. 'Catch-up' industrialisation was dependent on the availability of foreign loans at favourable interest rates (Chapter 5). The steep increase in interest rates following the 'Volcker Shock' led to a situation where the industrial profit of developing countries no longer exceeded the external debt burden. What was originally seen as a unique input by the creditor countries in order to instigate sustainable development in the debtor countries turned into long-term dependency mediated by the international credit and currency organisations. A range of developing countries had no option but to agree to restructuring and adjustment plans with the International Monetary Fund (IMF) and World Bank (WB). These plans included wage freezes and cutback policies, especially in the area of social services, in order to ensure that the debtor nations were creditworthy (Hall and Midgley, 2004). The debt trap in which many of the developing countries found themselves in the 1980s indeed signalled the end of the postwar development strategy of industrial import substitution. Although initially belittled, the neoliberal reforms of the 'Chicago boys' in Chile proved comparatively successful in these new circumstances. Chile's growth strategy, which achieved an average GDP growth rate of 6.4 per cent during the 1990s (Table 9.3), barely changed when the country returned to democracy in 1989. This was seen by many as the new model for catch-up development; most South American countries followed it and privatised their state-owned companies, abolished import restrictions and trimmed budget deficits. Even though annual GDP growth rates in other South American countries were not as impressive as those in Chile (the exception being Peru in the 2000s), this type of neoliberalism – one that was run by an authoritarian state – which had made Chile an attractive location for transnational capital, achieved recognition well beyond the continent of Latin America. Examples from South-East Asia – South Korea, Taiwan and Singapore in particular – provided additional evidence that dictatorship and neoliberalism were not incompatible (Harvey, 2006). These countries featured annual GDP growth rates between 5.3 and 7.6 per cent in the 1990s and between 2.9 and 4.6 per cent in the 2000s. But India's GDP growth also accelerated in the 2000s, as did that of several African states.

Given the Chilean 'success model' of a combination of authoritarianism and neoliberalism, China's rise to an economic power of the first degree is of special significance. This rise occurred within the context of the coincidence of the domestic socio-economic reform programme announced by Deng Xiaoping in 1978 with the turn towards neoliberal solutions in the US and the UK. Harvey (2006, p. 35) points to the fact that China's extraordinary economic performance, with annual GDP growth rates of nearly 10 per cent over the last three decades (Table 9.3), would have been more difficult 'had not the turn towards neo-liberal policies on the world stage opened up a space for China's tumultuous entry and incorporation into the world market'. In the course of the reform process, China opened up to foreign capital in order to restructure much of its manufacturing sector. In many cases, transnational corporations took over less productive state enterprises, admittedly often without also taking over company-related pension and welfare schemes and without granting workers co-determination rights. Since an authoritarian industrial relations regime allowed transnational corporations to structure and dominate the work process and to produce with much lower wage costs than in the West, China became an attractive location for FDI (Table 9.4). This especially applied to those sectors where industrial mass production predominated, sectors located in the Atlantic space during Fordism. However, in contrast to what

Table 9.3 GDP growth in the developing world: 1980–2009

	1980–89	1990–99	2000–09
South America			
Argentina	-0.9	4.2	3.2
Brazil	3.0	1.7	3.2
Chile	3.6	6.4	3.7
Colombia	3.4	2.9	3.9
Mexico	2.4	3.4	1.9
Peru	0.6	3.3	5.2
Asia			
Bangladesh	3.3	5.3	5.8
China	9.8	9.9	9.8
Hong Kong	7.4	3.6	4.1
India	5.4	4.8	7.0
Singapore	7.5	7.6	4.6
South Korea	7.7	6.3	4.3
Pakistan	6.6	3.9	4.5
Taiwan	8.2	6.5	2.9
Thailand	7.2	5.3	4.0
Africa			
Ghana	2.0	4.4	5.3
Kenya	4.6	2.1	3.7
Morocco	3.9	2.8	4.8
Nigeria	1.8	2.6	8.3
South Africa	2.2	1.4	3.5
Tanzania	3.0	3.1	6.6

Note: Gross domestic product in constant prices; annual percentage change.
Source: IMF

happened in many other developing countries, China's strategy of world market integration was not built exclusively upon the exploitation of cheap labour, but was also accompanied by large-scale investment into roads, railways, education and research, in a manner reminiscent of the US New Deal of the 1930s.[2] This could well prepare the second stage of China's world market integration, a stage based on product diversification and skilled (and correspondingly better paid) labour. Whether this will be accompanied by a democratisation of economy and society is a different question.

United Nations Conference on Trade and Development (UNCTAD) data on FDI inward stocks by region and country between 1990 and 2008 (Table 9.4) demonstrate not only that these stocks increased

Table 9.4   FDI inward stock by region and economy: 1990–2008 ($million)

Region/Economy			FDI inward stock		
			1990	2000	2008
World			**1,942,207**	**5,757,360**	**14,909,289**
	Developed economies		**1,412,605**	**3,960,321**	**10,212,893**
		EU	761,897	2,163,354	6,431,893
		Czech Republic	1363	21,644	114,369
		France	97,814	259,775	991,377
		Germany	111,231	271,611	700,471
		Netherlands	68,731	243,733	644,598
		Poland	109	34,227	161,406
		Spain	65,916	156,348	634,788
		Sweden	12,636	93,995	253,502
		UK	203,905	438,631	982,877
		Other developed Europe	**47,045**	**118,209**	**500,632**
		North America	**507,754**	**1,469,583**	**2,691,160**
		Canada	112,843	212,716	412,268
		US	394,911	1,256,867	2,278,892
		Other developed economies	**95,908**	**209,175**	**589,207**
		Australia	73,644	111,139	272,174
		Japan	9,850	50,322	203,372
	Developing and transition economies		**529,602**	**1,797,039**	**4,696,396**
		Africa	**60,635**	**154,244**	**510,511**
		South Africa	9207	43,462	119,392
		South America and the Caribbean	**110,547**	**502,487**	**1,181,615**
		Brazil	37,143	122,350	287,697
		Chile	16,107	45,753	100,989
		Cuba	2	74	185
		Mexico	22,424	97,170	294,680
		Asia and Oceania	**358,412**	**1,079,435**	**2,583,855**
		China	20,691	193,348	378,083
		Hong Kong	201,653	455,469	835,764
		India	1,657	17,517	123,288
		Singapore	30,468	110,570	326,142
		Turkey	751	19,204	69,420
		South-East Europe and CIS	**9**	**60,872**	**420,414**
		Russian Federation	–	32,204	213,734

Source: UNCTAD (2009, pp. 251–3)

significantly in total, but that the developing economies in Africa, Asia and South America received a greater share of FDI. In 1990, $1,412.6 billion or 72.7 per cent of the global FDI stock was allotted to the developed countries, while $529.6 or 27.3 per cent was invested in the developing countries. By 2008, the developing countries' share had increased to 31.5 per cent ($4,696.4 billion in absolute terms) and that of the developed countries had decreased to 68.5 per cent ($10,212.9 billion). In absolute terms, the FDI stock in developing countries grew almost by factor 10. In 2008, about one-fourth of the total FDI invested in developing countries went to China and Hong Kong. In China, the FDI grew by factor 19 in the period 1990–2008. In that year, the combined FDI inward stock of China and Hong Kong of $1,213.8 billion was only exceeded by the US. Further developing and transition economies that featured above-average FDI growth rates include Russia and the Community of Independent States (CIS), where the FDI inward stock grew from just over zero to $420.4 billion, and also Mexico and Cuba (the latter, like the CIS, had a very low point of departure). In Africa, growth in FDI proceeded at about the same speed as worldwide. In total, the – albeit slight – tendency towards geographic dispersion of FDI means that the attractiveness of developing countries as destinations of investment has increased. Conversely, transnational capital has become more important for economic and social decision-making processes in the developing countries (World Bank, 2007, p. 37).

Deregulation and world market openings across the globe provided enormous new investment opportunities for hedge funds and similar finance institutions, which enjoyed extraordinary growth rates when compared to the real economy. However, this further raised profit expectations beyond the limits of what could realistically be produced in the real economy, regardless of the conditions of the production process. From Chile's privatisation and commercialisation of the pension and welfare system through Margaret Thatcher's sale of large sections of social housing in Britain to the opening up of China and the former Soviet Union to the world market – which made available for capital accumulation hitherto unavailable assets and resources – all these processes turned goods and social relations formerly regulated collectively into individually owned wealth and private property rights. Harvey suggests using the concept of 'accumulation by dispossession' as a theoretical analysis of this phenomenon. Following Rosa Luxemburg, this 'concerns the relations between capitalism and non-capitalist modes of production'. The 'predominant methods' of this type of accumulation are 'colonial policy, an international loan system [...] and war' (Luxemburg, 1968, cited in Harvey, 2005, p. 137). Marx treated the original and 'primitive' accumulation as a historical process of commodification and privatisation of previously common land by using the example of the Scottish Highland Clearances, when the peasant population was forcefully expelled and collective property rights were changed to exclusive private property rights (Marx, 1961, Chapter 24). While most scholars regarded this process as being historically complete, Harvey (2005, pp. 145–6) maintains that primitive accumulation has 'remained powerfully present within capitalism's historical geography up until now'. I would agree that the ongoing processes of corporatisation, privatisation and commodification of previously public assets, from water and public utilities to social housing, telecommunications, transportation and universities, can be interpreted as a new way of 'enclosing the commons'. A further analogy with the Highland Clearances is that a strong state is normally necessary and frequently used 'to force such processes through even against popular will' (p. 146).

Accumulation by dispossession often has disastrous material and ecological results. The transition from common land to private property and the potential disappearance of the tropical rainforest is a prime example. The Amazon rainforest has frequently been described as the 'lungs of the planet', since it provides the world with the essential service of functioning as a CO_2 sink by continuously turning CO_2 into oxygen. More than 20 per cent of the world oxygen is produced in the Amazon rainforest. It is not only one of the most important counter-mechanisms to global warming, but also provides the natural environment for one half of the world's estimated 10 million species of plants, animals and insects, as well as for about 200,000 human beings.[3] Furthermore, 80 per cent of the developed world's diet originates in the tropical rainforest: fruit like avocados, coconuts, figs and mangos, vegetables like corn, potatoes and rice, and spices such as black pepper, cinnamon, turmeric and ginger. The coverage of the earth's land surface by rainforests has receded from 14 per cent before the Industrial Revolution to a mere 6 per cent in 2009, and UN experts fear that the last remaining rainforests could be destroyed in less than 40 years.[4] Rainforests are being cut down because multinational corporations, landowners and national governments perceive their exchange value as being only that of timber. In contrast, remedying climate change and maintaining the world's biodiversity is a use value to the world that costs nothing and is thus ignored by profit calculations of businesses. Since rainforests are comparatively 'worthless' from the standpoint of capital valorisation, they are being replaced by more profitable farms and ranches. The rainforests' territory is being commodified to the economic benefit of multinational corporations.

10
A Worldwide Consumption Norm (Based on Debt) and the Financial Crisis

The steady advancement of real wages formed the structural background for mass consumption, which itself paralleled the rise in profits, and thus constituted a fundamental pillar of the Fordist growth model. Like gross domestic product (GDP) growth (Table 8.2) and the wage share (Table 8.4), the growth of real wages diminished greatly throughout the entire Atlantic space beginning with the 1980s (Table 10.1). This development culminated in the 2000s, when real wage growth almost stagnated. It was well below 1 per cent in EU 27, the US and Japan. It was only in the Republic of Ireland, Hungary and the Czech Republic that growth rates of real wages of over 2 per cent were recorded during this decade.

If real wages stagnate and demand levels are still to be maintained, private consumption must be financed and fuelled by debt. Many economists and policymakers were of the opinion that the capitalisation not only of corporate ownership, but also of household assets, would make up for the lack of growth in the real economy. Rising levels of debt were regarded as unproblematic as long as the gains from financial investments, especially in the housing market, covered the debts. Consequently, households were given more access to credit, not only in the form of mortgages but also in the form of consumer loans, credit cards and bank accounts with overdraft provisions.[1] Table 10.2 presents Organisation of Economic Cooperation and Development (OECD) data on household debt as percentage of disposable income in major Western countries. While debt ratios rose in all countries but Germany from 1995 to 2005, in Spain, Ireland, the Netherlands and the UK household debt grew especially fast. Foster and Magdoff (2009, p. 29) point out that this co-occurrence of stagnating real wages and soaring consumption

Table 10.1 Real wages in selected European countries, US and Japan: 1961–2010

	1961–1970	1971–1980	1981–1990	1991–2000	2001–2010
Austria	4.6	3.5	1.4	1.7	0.7
Belgium	4.3	5.1	1.0	1.6	0.5
Denmark	4.6	3.0	0.5	2.5	0.4
Finland	3.7	3.4	2.4	1.1	1.9
France	5.2	3.7	0.9	1.2	0.5
Germany*	4.6	3.0	0.5	2.5	0.4
Italy	5.9	3.1	1.1	0.2	0.2
Netherlands	5.2	2.9	0.6	1.1	1.2
Sweden	4.1	1.6	0.6	2.1	1.0
UK	2.7	1.8	1.9	1.8	1.0
Ireland	4.1	4.2	2.1	1.2	2.4
Portugal	7.0	4.0	1.5	3.1	0.9
Spain	7.1	4.6	0.9	0.7	0.6
EU 15	4.5	3.0	0.8	1.4	0.6
Poland	–	–	–	5.4	1.1
Czech Republic	–	–	–	2.6	3.3
Hungary	–	–	–	-0.8	2.1
EU 27	–	–	–	1.6	0.6
US	2.4	1.1	1.0	1.8	0.7
Japan	7.3	4.9	2.4	1.0	0.5

Note: Real compensation per employee, deflator GDP; annual percentage change.
*1961–91: West Germany
Source: EU Commission: *European Economy*, Autumn 2009: Statistical Annex

Table 10.2 Household debt as percentage of disposable income in selected countries

	1995	2000	2005
Denmark	188	236	260
Germany	97	111	107
Spain	59	83	107
France	66	78	89
Ireland	–	81	141
Netherlands	113	175	246
Finland	64	66	89
Sweden	90	107	134
UK	106	118	159
US	93	107	135
Japan	113	136	132

Source: Girourard et al. (2006, p. 9)

based on consumer debt was made possible by 'historically low interest rates' in all major financial centres. This facilitated the servicing of debt in the 1990s and much of the 2000s. A large part of household debts were in the form of mortgages. As long as house prices went up, people – not only in the US and the UK, but also in many other Western countries such as Spain or Ireland – were encouraged to purchase new homes and indebt themselves even more by obtaining new mortgages on existing homes on the basis of the estimated value of their houses. In addition, new kinds of mortgages were developed for people who could not actually afford to buy a property. In the 'subprime' market segment, comparatively low interest rates were charged for the first few years before rates became adjustable. The result of this was an era of hyperspeculation and 'Ponzi finance' (Minsky, 1982).[2] The real-estate business boomed and many speculators enjoyed extraordinary profit rates by purchasing houses in order to resell at higher prices later. Homeowners, on their part, began to view happily the rapid increase in the value of their homes as 'natural and permanent, and took advantage of low interest rates to refinance and withdraw cash value from their homes' (Foster and Magdoff, 2009, p. 97).

Housing bubble and credit crunch

In order to limit the risks of mortgage default and to stabilise the new financial edifice, banks began to pool individual mortgage loans – prime and subprime mortgages – by using the cash flow provided by these loans to generate residential mortgage-backed securities. Subsequently, these 'securitised loans' were themselves repackaged in the form of 'collateralised mortgage obligations' (CMOs). This pool of diverse individual mortgages consisted of different 'tranches', some of which received the status of 'first claim' on the mortgagees' payments. Only 'once these claims were satisfied was money sent to less senior shares' (Krugman, 2009, pp. 149–50). In the 1990s, banks began to create 'collaterilised debt obligations' (CDOs), which mixed together low-risk, middle-risk and high-risk (subprime) mortgages with other types of debt. These were subsequently sold to investors, who assumed that 'geographic and sector dispersion of the loan portfolio and the "slicing and dicing" of risk would convert all but the very lowest of the tranches of these investment vehicles into safe bets' (Foster and Magdoff, 2009, p. 95). This assumption was usually confirmed by the three major US rating agencies Standard & Poor's, Moody's and Fitch, which tended to award the highest tranche of these CDOs the best possible ratings ('AAA').

The Federal Reserve (FED) further fuelled the market by suggesting that homebuyers were wasting money by buying fixed-term mortgages instead of adjustable mortgages (Baker, 2008, p. 74). Standards in lending policies were partially lax, due to the general belief in that house prices would rise permanently, and partially because lenders did not worry about the quality of their loans since they did not plan to keep them. Housebuyers were allocated loans that required low or no deposits, and with monthly payments that were actually beyond their ability to pay once the initially low, 'teaser' interest rate ceased to apply. The 'subprime' mortgage market grew accordingly and housebuilding boomed to meet rising demand.

The art of stabilising and expanding private consumption in a period of stagnating real incomes has always been dependent on rising house prices, since, on average, home equity decreases and mortgage debt service increases. The housing and household debt bubble originated from a calamitous combination of low interest rates, rapidly rising household debt and a worsening ratio of mortgages compared to downpayments. This was bound to burst at that moment when house prices started falling and households fell into the trap of negative equity. This is precisely what has happened since 2008. In his comparison of inflation-adjusted house prices in the US since 1895, Robert Shiller (2006) demonstrates that real house prices remained essentially unchanged in the century prior to 1995. By 2002, however, adjusted house prices had risen by almost 30 per cent. Paul Krugman (2009, p. 169) estimates that the overvaluation of the US housing market amounted to more than 50 per cent in 2006. For that year, he calculates the number of American homeowners with negative equity to be around 12 million. These were prime candidates for default and foreclosure when house prices started to fall in 2006. When houses became hard to sell, renegotiations of individual mortgages became more difficult, since subprime loans were normally not negotiated by the banks that held the loans. The original loans were usually quickly sold on to financial institutions, which, in turn, sliced and diced pools of mortgages into CDOs and sold them on to investors, creating a complex and dispersed structure of relations between creditors and debtors. The burst of the bubble was triggered by Bear Stearns' hedge funds' failure and accelerated by the bankruptcy of Lehman Brothers. This led to a credit crunch in the entire banking sector, when default risks and risk premiums shot up. Banks had financed their investments by heavy borrowing from other banks and finance institutions around the world and, due to the complexity and rapidity of developments, they found themselves in a situation where they 'didn't know whether what they owed to their depositors and bondholders exceeded the value of their assets' (Stiglitz, 2010, p. 2). Since they could not know the positions of other banks, 'the trust and confidence that underlie the banking system evaporated. Banks refused to lend each other – or demanded high interest rates to compensate for bearing a risk. Global credit markets began to melt down' (p. 10).

What started as an US American housing bubble soon turned into a credit crunch that reached Europe and other parts of the world, taking the form of a 'freefall' (Stiglitz, 2010) of the global economy. With regard to the EU, this transmission first occurred via bank losses. European banks had invested greatly in dubious or 'toxic' assets from US financial institutions, including Lehman Brothers. As a result of this interconnectedness, European banks had to write off losses of over $1,500 billion in 2009 alone (EuroMemorandum Group, 2009, p. 3). As in the US, European money markets experienced severe constraints in bank lending. The second transmission channel was worldwide trade, which fell by 33 per cent in the second quarter of 2009 compared to the previous year. Foreign direct investment (FDI) likewise decreased: a negative growth (-14.2 per cent) in FDI inflows featured in 2008, as compared to 2007 (Table 9.1). When imports of manufactured goods were cut back, countries with an export-dominated economy, such as Germany, were hit hardest. But exports also decreased in China and India. Countries in central Europe and the Baltic region had been running large current account deficits financed by borrowing on the international capital markets. These countries were left in a vulnerable position when this source of financing suddenly ran out (EuroMemorandum Group, 2009, p. 4). As a result, the world GDP was negative in 2009 for the first time since the Second World War. The social consequences were dramatic. The International Labour Organisation (2009) estimates that global unemployment increased by more than 50 million between 2008 and 2010 and that some 200 million workers were pushed into extreme poverty. In addition, many millions of people involuntarily worked short hours or part-time. Pensioners' income fell and home foreclosures became a mass phenomenon. A range of countries had to 'bail out' major banks and run up massive debts as a result. Some countries, such as Iceland or Greece, came close to state bankruptcy in the aftermath.

The expansion of Western consumption patterns

Even though the structural imbalances inherent to finance-driven capitalism were noticeable before the onset of the crisis, they were ignored and aggravated politically right up to the last minute.[3] On the contrary, as long as house prices kept rising, the resulting 'easy cash' for home-owners was perceived as a lasting and natural feature of the Western way of life. There were no financial boundaries to individual consumption, and so it became truly limitless as a result. While the Fordist mode of consumption was largely built upon the purchase of durable consumer goods, in the course of subsequent waves of commodified consumption more areas of life were restructured. Several of these followed developments in information and communication technologies such as the Internet. Entire new industries opened up in the 'entertainment' business, for example, as new generations of home computers, laptops and iPods were launched on the market at ever shorter intervals. 'Shopping', be it in malls or online, took on a new quality; in contrast to the postwar period, when many of the purchased goods had essential use values (for washing, clothing, mobility etc.), post-Fordist consumption patterns seemed to transcend the sphere of necessity. Recent sociological and anthropological analyses illustrate that the level of income that Western middle classes have at their disposal allows them a lifestyle that means they do not need to worry about material existential matters. However, since increases in income signal differences in authority, power, class and status – especially in highly unequal societies – they also advance one's position in the societal race for the positional goods that express and measure people's social standing via distinction (Chapter 3). Consumer researchers and social psychologists assume that we develop an attachment to the things we own, which causes us to perceive material possessions as part of our 'extended self' (Belk, 1988, cited in Jackson, 2009, p. 64). Some of these possessions are 'fleeting' (Jackson, 2009, p. 64), that is, 'they burn with novelty momentarily and are extinguished as suddenly something else attracts our attention'. Others are more durable, sometimes providing a 'sanctuary for our most treasured memories and feelings'. In an increasingly secular world, individual choices of what is 'sacred' and what is 'mundane' are usually no longer made in religious terms – as Durkheim (2001) observed a century ago – but upon the grounds of what Jackson calls the new materialism that functions as a substitute for religious consolation. It is precisely 'the cornucopia of material goods and its role in the continual re-invention of the self' that distinguishes the present consumer society from its predecessors. It is in contemporary circumstances that 'this wealth of material artefacts [has] been so deeply implicated in so many social and psychological processes' (Jackson, 2009, p. 64).

The popularisation of cultural practices that were formerly the privilege of the rich and powerful and the competition for positional goods via strategies of distinction also have a geographic and ecological dimension. Flying, for example, causes very high levels of CO_2 emissions, thus directly influencing climate change. In the 1960s, a flight to Mallorca was a special event, which couples might have enjoyed on their honeymoon, but not every year. Thanks to decreasing ticket prices, it has become 'normal' for an average Western wage-earner to fly to such a location for a long weekend. However, since more and more people are travelling to European destinations, the distinctive returns of these locations decrease, making it 'necessary' to travel even further away. The academic world, unfortunately, is no exception to this rule. Ongoing specialisation and international cooperation in most academic disciplines goes beyond the national or local sphere in many cases. In order to stay up to date in one's discipline, it is essential to share contacts with researchers from other parts of the world. The advance of new technologies, however, has made knowledge diffusion 'widely independent of the individual scientist's travel activity' (Mau, 2009, p. 69). The fact that personal meetings are nevertheless organised thus has little to do with knowledge diffusion and much more with the rules of the game in academia, where personal contacts constitute an important mechanism for creating and maintaining a reputation. Strategies of symbolic capital accumulation vis-à-vis one's peers include choosing faraway locations for conferences and similar academic events, and not local ones. The rule of thumb is that the further afield the location that an academic travels to (and gets his or her institution to pay for), the greater the symbolic gain measured in international esteem and excellence. In contrast, choosing not to attend conferences on other continents or reaching them by train or ferry smacks of the parochial and provincial. Needless to say, the ecological costs of such short trips by plane are rarely considered by travellers and their funding bodies. If they were, many of these trips would be unnecessary from the point of view of knowledge diffusion and plainly irresponsible from an egalitarian and sustainable perspective.[4]

The advertising industry thrives very well from the need of companies to sell consumer goods to people who often neither need nor have the means to purchase them but who nevertheless do so – on credit. A good example is the car industry, where around $10 billion was spent on advertising in 2003 in the US alone, and which has expanded immensely – compared to the modest beginnings of the automobile revolution in 1950, when there were 70 million cars, trucks and buses on the planet's roads. By 2025, the number of cars is expected to exceed 1 billion (Simms, 2005, p. 129).[5] By demonstrating the apparent indispensability of the car to our lives, marketing and advertising strategies build upon peoples' dreams and aspirations to get ahead as well as on their social and sexual identities. Advertisements uncritically reflect the contradictory relationship between people and nature in the context of capitalist society: 'Nature is big and powerful but you, in your car, are bigger and more powerful still' (p. 134). And, as in pornography, the male car purchaser has no wider obligations to society: 'This car, and women, won't answer back they'll just give you pleasure' (p. 136). The world portrayed in advertisements bears almost no resemblance to the 'actual condition of the product's manufacturing and consumption'. The monotonous tasks that dominate car manufacturing and the actual world of individual mobility, with its frequent traffic jams and the pollution it causes, never feature in advertising brochures or TV commercials. 'Released' from the real world of driving, the message is 'that you can take a short cut to finding purpose, meaning and wellbeing. Instead of taking a path of questioning and reflecting on how we live, the message is just go out and buy stuff to achieve fulfillment' (Simms, 2005, p. 137). Simms's analysis demonstrates that advertisement succeeds in making people believe that they 'could not live without a car', despite the fact that public transport is often available and that they may have used it regularly in their youth. The social use of cars in contemporary society has little to do with getting from A to B and features all the hallmarks of a physical and psychological addiction as a result of neoliberal ideology and, in particular, the brainwashing of customers by the advertising industry.

After import substitution had ended, the markets of many developing countries were flooded with new media images and foreign consumer goods as a result. 'Western' lifestyles were promoted in advertising and the media since the markets for consumer goods were not as saturated in the 'emerging economies' of China, India or Brazil, whose relative weight and general significance in the world economy has grown over the last decades (Chapter 9). Income levels rose for increasing percentages of the population; increasingly more people could afford consumption patterns previously limited to Western societies. Social structures and lifestyles evolved quickly as a result. A useful indicator for this transition is the enormous rise of the absolute number of members of the middle class in developing countries, which the World Bank (WB) (2007) estimated at 400 million people in 2004 and expects to rise to 1.2 billion people by 2030. In China, to take the most striking example, the middle class is predominantly located in the urban areas of the coastal mainland. Large cities are magnets for FDI, the service economy is concentrated there and significantly higher wages are paid there than in rural areas. Recent research indicates that consumption patterns of the emerging urban middle classes have started to differ qualitatively from those of the lower classes. The relative weight of necessity items such as food decreases in middle-class households, 'while discretionary spending (such as on recreation and education, transport and communication, housing and utilities, or household and personal items) is increasing' (Reusswig and Isensee, 2009, p. 132). As in developments in the US and in Western Europe in the 1950s and 1960s, more Chinese people are acquiring the classic 'Fordist' consumer durables such as televisions, private cars and their modern counterparts (PCs, iPods and the like) and, less often, a family property in suburban areas (Zhang et al., 2009, p. 145). This is being complemented by other cultural practices – also originally Western – such as tennis or golfing. At the same time, attitudes towards beauty contests, fashion shows, sexual behaviour and eating habits are changing and approximating Western ideas (Dittrich, 2009; Xu, 2007).

Due to their relative price reduction, consumer goods are increasingly available to middle-class households and the rapid speed at which consumer markets in Asia and (less so) Latin America are developing make it justifiable to speak of a 'second wave' of internal takeover of more and more areas of life by capitalism, following on from the first Fordist wave in the Atlantic space. Indeed, at no other point in time have so many people in the world participated in consumption patterns that used to be the privilege of elites. Developing countries, especially in Asia, are not just the 'factory of the world', due to the transferring and offshoring of industrial production from high labour-cost countries to cheap-labour cost locations. Just like Western Europe after the Second World War, developing countries have also become increasingly attractive as purchasers of 'Western' products. This was, in any case, the message of US President George Bush to China on a state visit to Australia: 'We certainly hope that China changes from a saving society to a consuming society. What we want is the government to provide more of a safety net so they (the Chinese people) start buying more US and Australian products' (quoted in Chua, 2009, pp. 112–13). Here a slowdown in consumption in the US was not proposed in order to reduce the enormous trade deficit between the US and China, but instead the Chinese are encouraged to consume more. If the world functioned according to Mr Bush's logic, the US would serve as a general 'consumption standard' for all countries. Indeed, there is evidence that this could happen. While the purchase of automobiles, for example, has reached a certain level of saturation in Atlantic core markets, the ownership of cars in the developing economies has been growing rapidly and is expected to rise further. Thus, if the level of car ownership in the US, where two out of three persons own a car, is to be taken as standard, '700 million Chinese and 600 million Indians will have to drive around in their own enclosed, private vehicles' (Ramachandra Guha, cited in Chua, 2009, p. 112). It is currently completely unclear where the natural resources are to come from for the necessary raw materials (such as steel and aluminium) to build all these cars, never mind the fossil fuels that would power them. As Chapter 11 demonstrates, the ecological consequences of this most recent chapter of capitalist expansion are indeed disastrous; but they are easily ignored.

11
The Globalisation of the Fossil Energy Regime

Part II has demonstrated that Fordism's industrial paradigm made use of methods of material throughput that would undermine economy and society for the future generations. The energy regime was dependent on the consumption of vast amounts of fossil-fuel resources; this was accompanied by the emission of enormous and growing amounts of CO_2 into the atmosphere, triggering the greenhouse effect. The question is whether the transition towards a finance-driven accumulation regime and the transnational relocation of production sites moderated or even overcame the fossil energy regime. The different policy strategies for a 'Green Deal' raise great hopes that a new growth period, where gross domestic product (GDP) growth is 'decoupled' from the emission of CO_2 and other environmentally harmful substances, could come about. In 'post-industrial' societies economic output becomes progressively less dependent on material throughput; thus the economy can continue to grow without facing any ecological limits. In short, Western production and consumption patterns as such would not need to change – only their energy base would. In order to test empirically the feasibility of a 'decoupling' of output and throughput, it is important to distinguish between absolute and relative decoupling. Jackson (2009, p. 48) defines relative decoupling as a 'decline in the ecological intensity per unit of economic output'. Resource impacts are diminished relatively to GDP. But these impacts do not necessarily decline absolutely and may continue to increase at a slower pace than GDP growth. Only a situation in which resource impacts decline in absolute terms counts as an absolute decoupling. In relation to climate change, it is of course the latter situation that is required, if the 50–85 per cent reduction in global carbon emissions that the International Panel on Climate Change (IPCC) (2007a) regards as necessary to stabilise atmospheric concentrations of CO_2 around or below 450 ppm by 2050 is to be achieved.

Relative and absolute decoupling

Relative decoupling is about producing goods more efficiently, about 'doing more with less: more economic activity with less environmental damage; more goods and services with fewer resource inputs and fewer emissions' (Jackson, 2009, p. 48). The hypothesis used by adherents of 'Green Deal' solutions is that since resource inputs represent a cost to producers, 'the profit motive should stimulate a continuing search for efficiency improvement in industry to reduce input costs' (ibid.). There is some evidence to support this hypothesis. According to the IPCC (2007b, p. 3), global energy intensity, the amount of primary energy necessary to produce one unit of the world's economic output, decreased by 33 per cent in the period 1970–2004. It is possible to attain a comparative view of the period 1980–2006 by using data provided by the US Energy Information Administration (EIA), which confirm IPCC calculations.

Overall, global energy intensity, measured as total primary energy consumption per US dollar of GDP, decreased in the period 1993–2006 (Table 11.1). However, this decline only amounted to about 8.5 per cent. Performances of regions and countries vary enormously. Within the Organisation of Economic Cooperation and Development (OECD), energy intensities declined by over 40 per cent between 1980 and 2006 in countries such as the UK, the US or Sweden, while in Australia and Spain this decline was much lower. Among the developing countries, the picture is even more heterogeneous. Huge countries such as South Africa, Mexico and Brazil displayed an increase in relation to their energy intensity, while most Asian countries and locations (the exception being Hong Kong), Cuba and, to a lesser extent, Chile displayed a decrease. The progress made in China is especially obvious, albeit from a very energy intensive point of departure. China, whose GDP and productive output have expanded enormously over the last three decades (Table 9.1), displayed an energy intensity nearly six times higher than the UK in 2006. It is worrying that, after many years of progress, China's energy intensity began to climb again in the period 2000–2006. Furthermore, Jackson (2009, pp. 49–50) demonstrates that improved resource efficiency was accompanied by declining CO_2 emission intensities. Between 1980 and 2006, the global carbon intensity of the GDP declined almost by a quarter, 'from just over 1 kilogram

Table 11.1 Energy intensity compared: Total primary energy consumption per $ of GDP: British thermal units (Btu) per (2000) US $

	1980	1987	1993	2000	2006
World	–	–	**13,533**	**12,416**	**12,385**
Europe	–	–	**9743**	**8888**	**8314**
Czech Republic	–	–	36,655	28,376	24,835
France	9757	8819	8850	8167	7767
Germany	–	–	8581	7510	7260
Netherlands	14,157	14,170	11,920	9859	9739
Poland	–	–	33,957	21,147	18,150
Spain	9472	9381	9185	9684	9196
Sweden	12,928	12,124	11,695	9247	7648
Turkey	9295	9948	10,827	11,859	10,968
UK	10,052	8570	8292	6692	5810
North America	**15,386**	**12,762**	**12,191**	**10,576**	**9411**
Canada	23,537	21,019	21,300	17,865	16,486
US	15,135	12,227	11,630	10,082	8841
Mexico	9955	11,099	10,661	9924	10,061
Africa	**17,272**	**20,147**	**20,459**	**19,934**	**18,265**
South Africa	27,963	34,435	34,355	34,549	30,550
Central and South America	**11,322**	**11,777**	**12,725**	**13,513**	**12,864**
Brazil	9421	10,110	12,009	13,271	12,561
Chile	14,301	12,462	12,392	13,458	12,964
Cuba	14,393	12,850	15,404	13,988	8450
Asia and Oceania	**12,413**	**12,328**	**12,361**	**12,553**	**14,484**
Australia	13,700	13,239	13,408	12,504	12,035
China	94,498	66,975	49,549	31,022	34,931
Hong Kong	4532	4402	4342	4764	5254
India	25,909	29,133	30,443	28,778	24,616
Japan	5410	4631	4527	4805	4467
Singapore	20,242	19,015	19,227	16,417	17,363
Eurasia	–	–	**122,139**	**107,140**	**79,166**
Russian Federation	–	–	(1994)		
			112,855	105,843	80,923

Source: EIA, 2008, Table E.1g
One Btu is equivalent to the amount of heat required to raise the temperature of one pound of water (at or near 39.2 degrees Fahrenheit) by one degree Fahrenheit. It is equal to 0.252 kilocalories or 1055.06 joules (http://www.businessdictionary.com/definition/British-thermal-unit-Btu.html).

of CO_2 per US dollar [...] to 770 grams of CO_2'. Again, the situation was uneven across the globe. While the OECD countries made steady progress, China's early improvements have been partially offset by increasing CO_2 intensity since 2000.

The world's energy supply

Jackson evaluates the progress made in terms of relative decoupling as 'faltering at best' since the efficiency with which the global economy uses fossil resources and generates CO_2 emissions improved in some locations, but not everywhere. Matters look worse still if we recall that 'relative' decoupling only measures the resource use and emissions per unit of economic output. If decoupling is to offer a way for the GDP to grow *and* for CO_2 emissions to fall, 'resource efficiencies must increase at least as fast as economic output does. And they must continue to improve as the economy grows' (Jackson, 2009, p. 50). Evidence of not only 'relative' but also 'absolute' decoupling in a growing global economy would, hence, include declining dependence on fossil fuels and an increased use of renewables. It is only under these conditions that global CO_2 emission levels would fall. The following tables provide information on the development of the world's total energy supply, on the contribution of renewables to this supply and on CO_2 emissions in relation to world regions and population growth. Taken together, the evidence implies that more than 30 years of post-Fordist restructuring – with the corresponding processes of financialisation, deregulation, opening up of ever more world regions to capital valorisation and more or less successful attempts at 'catch-up' industrialisation on the part of developing countries – have not been accompanied by any indication of independence from Fordism's fossil-fuel energy regime. On the contrary, the fact that the world's total primary energy supply, which stood at 6,115 Mtoe in 1973 (Table 7.1), almost doubled and amounted to 12,029 Mtoe in 2007 (Table 11.2), confirms the Jevons' Paradox (Chapter 2) once more: a decrease in companies' relative costs of fossil raw materials, which is indicated in the 'relative decoupling' of economic growth and resources in the Atlantic space, leads to an increase in their absolute demand.

The decision taken by multinational corporations to reorganise the production chains geographically – partly in order to avoid the wage, social security and environmental standards that had developed during the period of growth of Atlantic Fordism – has placed greater importance on developing countries as locations of industrial manufacturing.

Table 11.2 World's total primary energy supply 2007: Regional shares (total supply: 12029 Mtoe)

Regions	OECD	Africa	Latin America	Asia (excluding China)	China	Non-OECD Europe	Former USSR	Middle East	World Marine Bunkers
Shares (%)	45.7	5.2	4.6	11.4	16.4	0.9	8.5	4.6	2.7

Source: IEA (2009, p. 8)

In addition, relatively successful world-market integration strategies on the part of countries such as China and other Asian and South American countries, and the simultaneous collapse of 'real existing socialism', with the subsequent deindustrialisation of the economies of Russia and Eastern Europe, resulted in a shift in the regional distribution of the world's energy supply. Progress in energy intensity and efficiency in the Western countries also contributed to the fact that the OECD's share of the world's energy supply receded from over 60 per cent in 1973 to 45.7 per cent in 2007, while that of China rose from 7.0 to 16.4 per cent (Tables 7.1 and 11.2). The shares of the rest of Asia, Africa and South America also increased, while that of the former countries of the USSR decreased.

Not only has the world's energy use not receded or even remained stable since Fordism's peak, this supply has in fact continued to be met largely by fossil fuels (Table 11.3). Over four-fifths of the world's energy supply in 2007 came from fossil-fuel resources such as coal/peat, oil and gas. This was about 5 per cent less than in 1973 (Table 7.2), with gas and coal playing a larger role than oil. However, the 5 per cent increase in non-fossil-fuel energy sources was due only to an increase in the use of nuclear power – from 0.9 per cent of the world's energy supply in 1973 to 5.9 per cent in 2007. In contrast, the percentage of combustible renewables decreased from 10.6 per cent to 9.8 per cent, while that of hydropower slightly increased. The fact that the transition from a Fordist accumulation regime towards a finance-driven one has not been accompanied by a modification of the energy regime is further confirmed by a comparative analysis on the contribution of renewables to the total energy supply of Western countries (Table 11.4). In all OECD countries there was weak growth from 4.8 per cent in 1971 (Table 7.3) to 6.7 per cent in 2007, while in the European Union (EU) 27 member states renewables accounted for 7.3 per cent of the energy supply. This

Table 11.3 World's total primary energy supply by fuel shares 2007 (total supply: 12029 Mtoe)

Fuel	Coal/peat	Oil	Gas	Nuclear	Hydro	Combustible renewables and waste	Other*
Shares	26.6	34.0	20.9	5.9	2.2	9.8	0.7

Source: IEA (2009, p. 6)
*Includes geothermal, solar, wind, heat, etc.

Table 11.4 Contribution of renewables to energy supply (percentage of total primary energy supply)

Country	1971	1976	1981	1987	1992	1997	2002	2007
Czech Republic	0.2	0.3	0.4	0.4	1.5	1.7	2.5	4.9
Denmark	1.7	2.0	4.3	5.7	7.3	8.4	12.5	16.8
Finland	27.2	18.8	20.0	17.5	19.7	21.2	23.0	23.2
France	8.5	7.3	7.7	8.1	8.1	7.1	6.3	6.9
Germany	1.2	1.3	1.8	1.6	1.9	2.5	3.8	7.2
Hungary	2.9	2.2	2.0	1.6	1.9	2.5	3.8	5.1
Ireland	0.6	0.7	0.8	0.6	1.6	1.4	1.7	2.9
Netherlands		0.3	0.3	0.2	1.4	2.1	2.7	3.6
Norway	40.2	46.5	45.4	48.9	48.9	43.6	50.1	50.4
Poland	1.6	1.3	1.5	1.9	2.3	4.3	5.2	5.4
Portugal	18.9	12.4	11.4	14.3	15.3	17.4	13.8	18.0
Slovak Republic	2.4	2.0	1.7	1.5	1.6	3.9	4.2	5.4
Spain	6.4	2.9	3.2	3.8	5.3	6.4	5.4	7.2
Sweden	20.4	19.4	21.8	24.0	26.4	27.1	25.8	30.0
UK	0.1	0.2	0.2	0.2	0.7	0.9	1.4	2.3
US	3.7	3.9	5.1	5.6	5.6	5.2	4.3	5.0
EU 27					4.9	5.7	6.1	7.3 (2006)
OECD	4.8	4.7	5.4	6.1	6.2	6.2	5.9	6.7

Source: OECD: Economic, Environmental and Social Statistics

was slightly above the US, whose share has remained around 5 per cent since the 1980s. From a comparative point of view, it is worth noting that the highest percentages of renewables within the total of primary energy supplies were measured in the Nordic countries: in 2007, this percentage was 50.7 per cent in Norway, 30.0 per cent in Sweden, 23.2 per cent in Finland and 16.8 per cent in Denmark. Not only do these countries have a tradition of coordinated and negotiated indus-trial relations and a social–democratic welfare regime that has the best ranking with regard to combining economic growth and social cohesion (Koch, 2006a), but in terms of sustainability and use of renewable energy sources they also outperform other Western countries (the exception being Portugal, which was ranked above Denmark in 2007).

World population and carbon emissions

Table 11.5 shows the world population in relation to CO_2 emissions by region. The absolute number of people on earth rose from 3,917 million

Table 11.5 World population and CO_2 emissions by region 2007

Region	World Population (million)	Share of World Population	CO_2 Emissions[a] (Mt of CO_2)	Share of CO_2 Emissions[b]
World	6609	100	27,921	96.5
OECD	1185	17.9	13,001	44.9
Africa	958	14.5	882	3.1
South America	461	7.0	1016	3.5
Asia (excluding China)	2148	35.4	4287	14.8
China	1327	20.1	6071	21.0
Former USSR	284	4.3	2412	8.3
Eastern Europe (non OECD)	53	0.8	252	0.9

Sources: IEA (2009, p. 48)
[a] CO_2 emissions from fuel combustion only. Excludes international aviation and international marine bunkers, which made up 1041 Mt of CO_2.
[b] Excludes international aviation and international marine bunkers, which made up 3.5 per cent of CO_2 emissions.

in 1973 (Table 7.4) to 6,606 million in 2007. In the same period, the total annual amount of global CO_2 emissions almost doubled, going from 15,046 Mt to 27,921 Mt. In contrast to IPCC recommendations and far from representing any 'absolute decoupling' as defined by Jackson, the increase in CO_2 emissions is due not only to the growth in the world's population, but also – albeit to a smaller degree – to an increase in individual emissions. The CO_2 emissions of the average human being rose from 3.84 tonnes in 1973 (Table 7.4) to 4.22 tonnes in 2007.[1] Again, as in relation to the total energy supply, the shift in CO_2 emissions by regional percentages reflects the transition that has occurred in the international division of labour, and in particular the upgrading of Asia as a location for industrial production. Thanks to this relocation and progress in energy efficiency in industrial production and residential construction, the OECD's percentage of the world's CO_2 emissions went down from 65.8 per cent in 1973 (Table 7.4) to 44.9 per cent in 2007. However, this is still about two-and-a-half times more than the OECD's corresponding percentage of world population (17.9 per cent in 2007). Production and consumption patterns adopted in Western countries, hence, continue to be unsustainable with regard to their effect on climate change. Worse still, the accelerated export of

Western ideas of progress, welfare and lifestyle to the rest of the world has brought the planet closer to climate collapse. It is especially worrying that, since 1973, absolute CO_2 emissions increased in every single world region, with the exception of Eastern Europe; the industrial structure of these countries, established during the period of state socialism, was almost completely destroyed after the fall of the 'Iron Curtain'. The ex-USSR displayed only a small absolute increase in CO_2 emissions and a substantial drop in its relative percentage in global CO_2 emissions, from 14.4 per cent in 1973 to 8.3 per cent in 2007. In contrast, in non-Chinese Asia, the absolute and relative increase in CO_2 emissions was large. In 1973, this region's percentage of the world's CO_2 emissions amounted to just 4.0 per cent; by 2007, it had reached 14.8 per cent. However, if we consider that over one-third of the world's population lives in this region, then this percentage seems to be almost moderate.

China's growth in CO_2 emissions is spectacular. While the country's share of the world's population went down from 22.5 per cent in 1973 to 20.1 per cent in 2007, its percentage of CO_2 emissions increased from 5.7 to 21.0 per cent. An average Chinese citizen emits now slightly more than the average inhabitant of the world. In absolute terms, Chinese emissions grew by a factor of almost 7, which can be attributed to increasing industrial activity and lifestyle change. Comparatively low technological standards are reflected in relatively high levels of energy and carbon intensity. To meet rising demand for fossil fuels, China moved from self-sufficiency in oil production in the late 1980s to being the second-largest oil importer after the US. This had geopolitical implications since China began purchasing oil in the Sudan, central Asia and the Middle East to secure its oil supplies. At the same time, China also has vast low-grade supplies of coal, which has a relatively high sulphur content, making up more than 70 per cent of the country's primary energy consumption. Using coal for energy generation creates a range of environmental problems – climate change in particular. The partial adoption of a Western lifestyle has also increased fossil energy consumption. Consider the consumption levels of between 100 and 200 million people increasing significantly: 'the ecological footprint of China as a whole is growing – and by far move[s] beyond sustainable levels, reflecting a transition from industry-driven to lifestyle-driven environmental degradation' (Reusswig and Isensee, 2009, p. 134). Wei et al. (2007) attribute 26 per cent of China's total energy consumption and 30 per cent of CO_2 emissions in 2002 to lifestyle changes and to the economic activities that support these demands. Home energy use, food and cultural and recreation services were some of the most energy-hungry and CO_2 intensive practices. Carbon-intensive practices such as domestic and international flights and individual car ownership have increased. Nevertheless, viewing China as the main cause of the worsening of climate change, as many Western governments tried to do during the failed Copenhagen summit in December 2009, is missing the point and is hypocritical for two reasons: first, in contrast to the OECD countries, in 2005, China's percentage of global CO_2 emissions was only slightly above its percentage of world population; second, the increase in what is measured as being China's emissions is partially due to multinationals relocating huge sections of their industrial production from OECD locations to China, circumventing Western environmental standards in the process. In many cases, Western companies emit CO_2 from Chinese soil instead of emitting it from Europe or North America, as they did previously.

A comparison of CO_2 emissions by capita and country confirms this proposition (Table 11.6). Between 1975 and 2005, Chinese per capita

Table 11.6 CO_2 emissions 1975–2005: in tonnes per person and a country's percentage of the world's total CO_2 emissions

	1975	1980	1985	1990	1995	2000	2005
Australia	13.1t	14.3t	14.2t	15.4t	15.7t	17.9t	18.7t
	1.17%	1.15%	1.19%	1.25%	1.29%	1.45%	1.39%
Canada	16.5t	17.7t	15.9t	15.8t	16.0t	17.6t	17.3t
	2.46%	2.39%	2.19%	2.08%	2.14%	2.28%	2.03%
Denmark	10.7t	12.5t	12.0t	10.1t	11.4t	9.7t	9.0t
	0.35%	0.35%	0.33%	0.24%	0.27%	0.22%	0.18%
Finland	9.7t	11.8t	10.1t	11.2t	11.2t	10.6t	10.7t
	0.29%	0.31%	0.26%	0.26%	0.26%	0.23%	0.20%
Germany	12.5t	13.5t	13.2t	12.2t	11.0t	10.3t	10.0t
	6.30%	5.82%	5.43%	4.59%	4.08%	3.59%	3.01%
Hungary	7.1t	8.2t	8.0t	7.0t	5.8t	5.6t	5.9t
	0.48%	0.48%	0.45%	0.34%	0.27%	0.24%	0.22%
Ireland	6.9t	7.9t	7.7t	9.1t	9.4t	11.2t	11.0t
	0.14%	0.15%	0.14%	0.15%	0.15%	0.18%	0.17%
Netherlands	10.5t	11.0t	10.3t	10.7t	11.2t	11.0t	11.3t
	0.92%	0.85%	0.79%	0.76%	0.79%	0.74%	0.67%
Norway	6.4t	7.2t	7.0t	7.2t	8.0t	8.1t	8.4t
	0.17%	0.16%	0.15%	0.14%	0.16%	0.15%	0.14%
Poland	10.6t	12.0t	11.7t	9.3t	8.8t	7.8t	7.9t
	2.32%	2.35%	2.31%	1.68%	1.54%	1.27%	1.10%
Portugal	2.2t	2.8t	2.8t	4.4t	5.3t	6.4t	6.4t
	0.13%	0.15%	0.15%	0.20%	0.24%	0.28%	0.25%

Table 11.6 (Continued)

	1975	1980	1985	1990	1995	2000	2005
Russian Federation	13.2t	14.9t	16.2t	15.1t	10.8t	10.5t	11.0t
	11.41%	11.34%	12.41%	10.59%	7.29%	6.46%	5.70%
Spain	4.8t	5.4t	4.9t	5.7t	6.3t	7.6t	8.5t
	1.09%	1.12%	1.00%	1.05%	1.13%	1.29%	1.33%
Sweden	10.0t	9.1t	7.3t	6.4t	11.2t	6.2t	5.8t
	0.53%	0.41%	0.32%	0.26%	0.26%	0.23%	0.19%
UK	10.5t	10.3t	9.8t	9.9t	9.3t	9.0t	8.9t
	3.80%	3.18%	2.95%	2.67%	2.44%	2.24%	1.96%
EU 27	9.0t	9.7t	9.1t	8.9t	8.4t	8.2t	8.4t
	26.02%	24.49%	22.50%	19.89%	18.10%	16.74%	14.90%
US	20.4t	20.8t	19.4t	19.7t	19.5t	20.4t	19.9t
	28.27%	25.90%	24.49%	23.23%	23.52%	24.37%	21.40%
China	1.2t	1.5t	1.7t	2.0t	2.7t	2.6t	4.3t
	6.85%	7.92%	9.43%	10.96%	14.63%	14.09%	20.26%
Japan	7.8t	7.8t	7.5t	8.9t	9.5t	9.6t	9.8t
	5.63%	5.01%	4.83%	5.21%	5.38%	5.12%	4.54%
South Africa	8.6t	7.9t	7.4t	7.3t	7.2t	6.9t	7.2t
	1.37%	1.20%	1.24%	1.22%	1.28%	1.28%	1.22%
India	0.4t	0.4t	0.6t	0.7t	0.9t	1.0t	1.1t
	1.59%	1.67%	2.36%	2.94%	3.70%	4.31%	4.44%
Mexico	2.5t	16.0t	3.6t	3.7t	3.60t	3.9t	4.0t
	0.93%	1.31%	1.42%	1.44%	1.48%	1.62%	1.49%
Brazil	1.3t	1.6t	1.3t	1.4t	1.6t	1.9t	1.9t
	0.93%	1.05%	0.95%	0.99%	1.16%	1.39%	1.28%
Turkey	1.6t	1.8t	2.1t	2.5t	2.8t	3.3t	3.3t
	0.42%	0.43%	0.55%	0.67%	0.78&	0.93%	0.87%
Cuba	2.8t	3.0t	3.2t	2.7t	2.1t	2.3t	2.2t
	0.17%	0.16%	0.17%	0.14%	0.10%	0.11%	0.09%

Source: WRI

emissions grew by a factor of almost 4. According to IEA (2007) projections, China will reach a similar level of CO_2 emissions as all the EU countries combined by 2015. Between 2000 and 2005 alone, emissions rose from 2.6 to 4.3 tonnes per person. To date, however, per capita emissions in all Western countries were five times more than those of China. Throughout the observation period, individual emissions in the US were around 20 tonnes. Despite being considerably smaller in population than China and several other countries, the US was the world's biggest CO_2 emitter in 2005, with a percentage of the world's total CO_2 emissions of 21.4 per cent. In the EU 27, which occasionally lectures both the US and China for their respective contributions, or lack thereof, to climate change abatement, individual emissions decreased slightly, going from 9.0 tonnes in 1975 to 8.4 tonnes in 2005. This small reduction is due not only to improvements in energy efficiency, but also to the accession of Eastern Europe to the EU: its deindustrialisation after 1990 had a favourable influence on the EU's overall performance. The same applies to Germany, which reduced its per capita emissions after reunification and the phase-out of the German Democratic Republic's industry. In a comparative perspective, both within the EU and worldwide, Sweden's performance counts as 'best practice'. The country almost halved its per capita emissions, going from 10 tonnes in 1975 to 5.8 tonnes in 2005 and thereby reducing its percentage in worldwide CO_2 emissions from 0.53 to 0.19 per cent. Beyond Europe, per capita emissions increased in Japan, Mexico, Brazil, India and Turkey – with Mexico, Brazil and Turkey also increasing their relative shares in global emissions. Data provided by the Global Humanitarian Forum (2009, p. 3) indicate that the 50 least developed countries contribute less than 1 per cent to global CO_2 emissions, while they bear over nine-tenths of the climate change burden and the associated humanitarian disaster, namely '98 per cent of the seriously affected and 99 per cent of all deaths from weather-related disasters, along with over 90 per cent of the total economic losses'.

Cuba, alongside Sweden, is one of the few cases where both per capita emissions and the country's percentage of global CO_2 emissions decreased. Cuba is of particular interest, since the country continues to be disconnected from the mainstream world economy and its corresponding production and consumption regimes. Like many countries in Eastern Europe and like some ex-Soviet Asian countries, Cuba's economy went through a severe crisis following the breakup of the Soviet Union in 1990, particularly because of the loss of subsidised Soviet oil. But, while the Eastern European countries embarked on a transition towards some kind of capitalism, usually including the privatisation of state provision of health and social care, Cuba maintained its planned economy and its comparatively generous health-care and welfare system. The economy had to be renewed in relation to its energy basis. This was reflected in a decrease in energy intensity (Table 11.1). At the beginning of the so-called 'special period' after the fall of the Berlin wall, Cuban soils were infertile and incapable of feeding the population, while the US embargo limited the country's trading options. Somewhat paradoxically, recent studies demonstrate that the crisis led to significant health improvements, since during that period 'caloric intake was reduced by over a third' (Jackson, 2009, p. 44). The result was not starvation, but the fact that 'obesity was halved and [the] percentage of physically active adults more than doubled' (ibid.). There was also a significant decline in deaths caused by diabetes, coronary heart disease and strokes.[2] But equally important is the fact that the country successfully initiated the transition of its energy regime towards a sustainable one. Following a Soviet agricultural approach, Cuba had largely ruined its farmland with pesticides. Organic farming methods were adopted that gradually restored soil fertility and raised productivity. In addition, in large cities such as Havana, tracts of land were leased to farmers for no rent, as an incentive to produce food. A huge number of small agricultural entrepreneurs emerged, growing vegetables in cities and selling their surpluses at local farmers' markets. These changes were doubtlessly born out of an acute crisis situation and were implemented by a government that did not hesitate to use authoritarian methods against political opposition when it felt the need to do so. The results are nonetheless noteworthy. The World Wildlife Funds (WWF) measures countries' progress towards sustainable development by using the UN's Human Development Index (HDI) as an indicator of well-being, as well as the 'footprint' of countries as a measure of pressure on the biosphere. While the HDI is calculated using life expectancy, literacy and education as well as GDP growth, a 'footprint lower than 1.8 global hectares per person, the average biocapacity available per person on the planet', denotes sustainability at global level (WWF, 2006, p. 19). The result of the WWF's comparative measurement of all world regions and countries using these criteria is that 'no region, nor the world as a whole, met both criteria for sustainable development. Cuba alone did...'.

Part IV

The International Regulation of Climate Change or the Commodification of the Atmosphere

The historical development of markets and capital tends to rearrange communities that were previously isolated and to regroup their inhabitants according to new spatio-temporal structures. The spatial dimension of regulation is not one fixed object but a delicate structure, which is subject to rescaling processes in the course of which new, multi-scalar structures of state organisation, political authority and socio-economic regulation emerge (Boyer and Hollingsworth, 1997; Brenner, 2004). The notion of *spatio-temporal fixes* has been developed to reflect the fact that particular accumulation regimes correspond to certain scales of regulation or spatial boundaries, in which structural coherence is sought (Jessop, 2002, p. 21). The Fordist growth model, for example, and its spatio-temporal fix, which is based upon the nation-state, came under pressure not only through various processes of deregulation and re-regulation, but also through the corresponding de-scaling and rescaling processes that led to ongoing shifts in the sites, scales and modalities of the delivery of state activities. At the same time, foreign agents and institutions become more significant as sources of domestic policy ideas, policy design and implementation. To the extent that the increasingly transnational processes of capital accumulation require forms of regulation that extend beyond the borders and the capacities of individual states, governments attempt to create or strengthen international regulatory systems – in compensation, as it were, for the loss of ability to intervene at national level. Part IV analyses the framework of international climate regulation. Chapter 12 reviews the academic debate on multi-level governance and applies the results to the international climate negotiations known as 'the Kyoto Process'. Special emphasis is placed on global inequality, on associated difficulties in establishing trust between the different parties and on the difference, during negotiations, in interests and relative weight between the actors and institutions involved. Chapter 13 deals with the procedural terms of climate governance and outlines the theoretical origin and practical implementation of market-based approaches towards environmental economics and, in particular, the 'flexible mechanisms' that aim to bring about the necessary reductions in CO_2 emissions – those reductions that could avert life-threatening climate change. Chapter 14 points to the theoretical and empirical flaws of such market-based attempts and highlights the homology between the regulative imperatives of a finance-driven accumulation regime (discussed in Part III) and the flexible mechanisms adopted in climate policy.

12
Multinational Governance in an Unequal World: The Kyoto Process and the Actors Involved

The term 'governance' is often used to identify and theorise about evolving mixes of formally legal agreements between states and non-formal types of indirect governing, often via 'steering' and networking. Although governance approaches are increasingly deployed in transnational contexts, they began at the national level. During the crisis of Fordism at the latest, it became evident that a state's capacity to act was limited in domains such as economic, employment and social policies as well as in environmental protection. The influence of non-state actors in socio-economic regulation became greater in the situation of globalisation, and the idea of cooperation between public and non-public actors, for example in the form of 'public–private partnerships', gained both in importance and popularity. At the national level, this was accompanied by the concept of the 'cooperative state', which 'steers' the various actors who participate in governance networks via negotiations, cooperative conflict resolution and agreement. According to Renate Mayntz (2009, p. 164), the point of departure for governance theorists is not the description of a problem or a policy target but the analysis of existing institutions and the issue of their functions and of how well they fulfil them. Problems are seen as 'challenges' in relation to a theoretically constructed optimum situation, which at the same time serves as benchmark for institutional change. General normative goals such as 'sustainability' or, in our case, the reduction of greenhouse gas emissions by a certain percentage would thus be seen as an optimum situation against the background of which the current situation is assessed and addressed.

The advantages of horizontal and coordinative governance techniques, as compared to more hierarchical forms of government, have often been emphasised. First, 'governance' allows for the integration of different sources of knowledge, interests and norms and is seen as more flexible and adjustable than traditional forms of 'top-down' government (Scharpf, 1993). Second, transnationalisation processes such as European integration, as well as the multi-level process of climate negotiations, can be understood as emerging governance structures. Indeed the concept is valuable for the description of policies that do not resemble, or indeed have, a government but are instead characterised by 'institutional idiosyncrasy', 'vertical complexity' and 'processual ambiguity' (Leisink and Hyman, 2005, p. 277). State action is part of the focus of analysis, as are non-state actors and their capability to influence negotiations between states. However, there is no general agreement as to whether the pluralisation of decision-making processes leads to more democratic participation thanks to the inclusion of non-state actors or leads to increasing informality and the undermining of democratic institutions through unelected elites.[1] Third, Leisink and Hyman (2005, p. 278) show that the concept of 'governance' can be used to describe the interaction and often 'uncertain interdependence' between national and transnational authority. Fourth, in procedural terms, the concept provides a good depiction of the ways in which 'hard laws', which tended to regulate national spaces – especially in the postwar period – are supplemented and sometimes replaced by 'soft laws', often on transnational scales of regulation.

The disadvantages of governance and horizontal cooperation are less often discussed. Jachtenfuchs (2001) highlights the greatest disadvantage of the concept, pointing out that its focus has mainly been on the description of procedural processes, often to the neglect of issues of power relations. Similarly, Leisink and Hyman (2005, p. 279) criticise the tendency of governance theorists to fail 'to consider who has the power to define what are problems'. Neglecting or downplaying power is indeed a serious weakness of governance, since empirical research into governance processes at the national level indicates that power differentials and asymmetries between negotiating parties often translate into blocked decision-making and lead to compromises at the lowest common denominator and to the externalisation of costs towards third parties that are usually not part of negotiation processes. Often bargaining outcomes quite simply reflect the interests of the most powerful actors, while those who lack bargaining power at the outset of negotiations find themselves still lacking in bargaining power at the end. Further problems arise when applying and expanding governance at transnational and global levels. Mayntz (2009, pp. 170–1) points to the fact that the concept of the cooperative state, crucial at the national governance level, cannot be simply transferred to the global context since there is no 'world government' that could take on this role. Few of the institutional and structural preconditions for the establishment of governance structures as understood from the national context exist at global level. Not only is there no world government, political parties that could act globally do not exist either. There is, therefore, a discrepancy between the global reach of issues such as environmental damage and the limited abilities of nation-states to act. More narrowly defined, this discrepancy covers three levels of action (Mayntz, 2009, pp. 165–8). First, the level of originators of environmental damages, whose action is the cause of the problems; second, the level of the parties affected; and, third, the level of coping with the problem, which refers to those actors that attempt to provide a solution. Very often these three levels do not coincide in time and space, leaving the originators of a problem with the (often economic) benefits of their actions, while others bear the (often environmental) costs. While the originators could contribute to dealing with the problem but have no incentive to do so, those affected by it lack the necessary influence to bring about a solution. Finally, coping with problems at the international level encompasses a range of very diverse national organisations and institutions. Dealing with climate change, for example, necessitates global cooperation; however, the negotiation strategies of the actors involved very often focus on maximising their positional advantages and minimising their costs. Yet few public goods – goods in the general interest – are likely to be produced in situations where all the actors are motivated by the maximisation of advantages.

In relation to the issues of sustainability and climate change, Mayntz (2009, p. 169) argues that a cooperative solution to a problem is even more difficult to attain when the negative effects are not obvious to all. Such issues are of a diffuse and abstract nature, making it difficult for individuals to perceive and understand them. They are also complex, and their high level of uncertainty hinders both assessment and reaction. Since it is often hard to identify individual perpetrators, a range of resistance strategies against regulation emerge that would dismantle the advantages of the polluters. In environmental conflicts, then, vague and diverse public interests face the clear-cut and powerful material interests of individual polluters. Free-riding and resistance to environmental protection legislation are facilitated at the international level, where none of the participating actors and organisations can simply push through regulation or policies, leading on their own to more sustainable ways of running the economy. For this to happen, both the social forces that would benefit from greater sustainability and those who perceive themselves as losing out would all need to cooperate. Mayntz (2009, p. 175) is outspoken about the fact that such cooperation implies that corporations and their lobbies would need to reduce their profit expectations – which, under the auspices of finance-driven capitalism, is unlikely to happen. Similarly, in relation to the willingness of national governments to implement environmental protection, Mayntz argues that, in international negotiations, governments have mostly followed short-term national competitive interests rather than long-term environmental goals. Mayntz is of the opinion that convincing state actors to work towards sustainability is just as hard as persuading representatives of corporate interests to minimise their profit expectations. Thus, one of the main ingredients required for governance structures to function, which was stressed by Emile Durkheim (1997) more than a century ago, is largely absent: trust – or willingness, on the part of the negotiating partners, to embrace honest behaviour and reciprocity, which would increase and improve communication and information between partners. Trust, furthermore, 'reduces uncertainty and transaction costs, enhances the credibility of commitments, makes defection more costly, creates stable expectations, and ultimately promotes cooperative solutions' (Roberts and Parks, 2006, p. 41).

In international climate change negotiations, the production of trust is further complicated by the existence of three kinds of inequality between developed and developing countries; these are discussed by Roberts and Parks (2006). First, there is an 'inequality in responsibility' for climate change. Chapters 7 and 11 have demonstrated how developed countries are responsible for a much greater percentage of CO_2 emissions than their proportion of the world's population. This is complemented by an 'inequality in vulnerability' to climate change. While some populations, especially in Africa, Asia and South America, face disasters such as droughts, floods and storms, some of the largest contributors to global warming might even gain from the effects of climate change over the short term. As a corollary, there is an 'inequality in perception', which tends to translate into different and often opposed negotiation strategies at climate summits. Roberts and Parks (2006, p. 44) take the example of the 'cost' of climate change, which rich nations normally assess 'in terms of dollars, while poor nations tend to view the costs through the prism of lives lost and livelihoods destroyed'. I agree with these two authors that the lack of success in global climate negotiations has been exacerbated by the lack of trust caused by the continued existence of massive inequalities between the developed and the developing world. These will now be examined in greater detail.

Global income inequality

There are basically three different modes of comparing income inequalities across the globe. First, there are unweighted country comparisons – in which, like in the General Assembly of the UN, each country has the same weight (Luxembourg just as much as China); second, there are weighted country comparisons, which consider different levels of population (in our imaginary UN assembly, China would now have significantly more votes than Luxembourg); and, third, income inequality is calculated and compared at the household or individual level, without considering the country of origin. In order to measure international income inequalities, the International Comparison Project has collated information on relative price levels in different countries, which is used to calculate the purchasing power parity (PPP) exchange rates (Milanovic, 2005, pp. 12–19). The most frequently used indicator of inequality is the Gini coefficient, which maps inequality on a scale from 0 (all compared individuals or households dispose of exactly the same income) to 100 (the entire available income is monopolised by one person or household). In what follows, I build upon Branko Milanovic's pioneering study of 2005, which analyses World Bank data in relation to the three modes of income comparision presented above and over an extended period of time (Table 12.1).

Beginning with the unweighted country comparision, the Gini coefficient grew slowly but steadily between the 1950s and the 1980s.[2] During the period of 1988 to 2000, this growth was more rapid. Milanovic (2005, p. 44), whose set of data enables the breakdown of the World Bank data by continents and countries, explains the slower growth

Table 12.1 Global income inequality (Gini coefficients) 1952–2000

	1952	1960	1978	1988	1993	2000
Unweighted country comparison	45, 1	46, 3	46, 7	49, 8	52, 9	54, 3
Weighted country comparison	56, 9	54, 5	54, 4	52, 9	51, 6	50, 2
Households worldwide (PPP dollars)	–	–	–	62, 2	65, 3	64, 1
World population (millions)	2511	3025	4414	4900	5300	5845

Source: Milanovic (2005, p. 142)

of inequality during the first period through the relatively successful catch-up development of Eastern Europe, South America and, to a lesser extent, Africa. He attributes the faster growth of global income inequality in the 1990s to recession in South America after the debt crisis and to transition towards capitalism in Eastern Europe and in the former Soviet Union. Further, Milanovic (2005, p. 78) demonstrates a reduction in mobility by arranging countries according to a scale that divides GDP per capita growth into four groups: rich countries (above all, Western Europe, North America and Oceania); 'contender' countries whose GDP per capita is no more than one-third below that of the poorest country of the first group – these have a realistic chance of joining the club of the rich countries within a generation or two; the 'Third World', which comprises countries whose GDP per capita levels represent between one- and two-thirds of the level in the poorest country of the first group; and, finally, the 'Fourth World', composed of countries whose GDP per capita is less than one-third of the level of the poorest among the rich countries. In 1960, there were 41 non-Western countries that either belonged already to the richest group or were contenders for joining it (among them 16 South American countries). In 2000, however, there were only 17 such countries (5 of them South American). The club of rich countries comprised mainly of Western countries, while the number of non-Western countries in that group was reduced to nine. While the West 'reinforced its control of the top, being an African, country became synonymous with being very poor, much more so than probably ever in history' (Milanovic, 2005, p. 78). The African countries that had been contenders in 1960 were now relegated to the Third World, and '*all* African countries that were part of the Third World dropped to the level of the Fourth' (Milanovic, 2005, p. 78). At the end of the twentieth century, the likelihood of an African being part of the poorest group was 80 per cent. Only Asian countries were upwardly mobile towards the two richest groups: Taiwan, Singapore, Hong Kong, South Korea and Malaysia. Botswana and Egypt managed to move from the fourth into the third income group.

At first sight, the weighted country comparison indicates a trend in the opposite direction. The Gini coefficient does not increase but decreases – especially after 1978. However, Milanovic's breakdown of the aggregated data according to states and provinces (Milanovic, 2005, pp. 85–100) demonstrates that the Gini coefficient does not decrease but in fact increases – when China is excluded from the analysis. Indeed, the fall of the Gini coefficient coincides with the implementation of the Chinese economic reform policy. The latter appears to have contributed to the reduction of the income gap between China as a whole and the richest countries of the West.[3] In 1980, all Chinese provinces had a GDP per capita below the world's mean. In 2000, there were already five provinces with a GDP per capita above the global mean, while other provinces had reduced their distance. Milanovic (2005, pp. 99–100) concludes that 'growing interregional inequality in China and India has a discernible and positive effect on world inequality'.[4] If household incomes are compared globally, that is, independently of their country of origin, then income disparities become significantly larger.[5] The Gini coefficient then ranges between levels as high as 62.2 in 1988, rising to 65.3 by 1993 and slightly falling to 64.1 by 1998. Milanovic (2005, pp. 113–114) mentions three factors that explain both developments. First, incomes in rural but populous Asian countries rose more slowly when compared to incomes in Organisation of Economic Cooperation and Development (OECD) countries during the first period, but they caught up after 1993. Second, income differences between urban and rural areas within China and, to a lesser extent, India increased rapidly in the first period and decreased after 1993. Third, the relative reduction in the world's middle class, above all as consequence of political and economic developments in South America and Eastern Europe, led to an increase in income inequality during the first period. In the 1990s, however, real incomes, especially in Eastern Europe, started to catch up again.

Despite the recent fall in the world's Gini coefficient, global income distribution continues to take the form of a concentration both at the top and the bottom, while the percentage of the 'middle class' is still small. At the end of the twentieth century, 70 per cent of the world's population lived in countries whose GDP per capita was less than PPP $5,000 (Milanovic, 2005, p. 128). While 12 per cent lived in countries with GDP per capita levels between PPP $5,000 and $8,000, merely 4 per cent of the world's population lived in what Milanovic labels the 'broad middle income range' (countries with a GDP per capita between PPP $8,000 and $20,000). The remaining 14 per cent of the world's population lived in the 'rich' countries with a GDP per capita of above PPP $20,000. Within the richest group, there were likewise huge differences: 'The top 10 percent of the US population has an aggregate income equal to income of the poorest 43 percent of people in the world, or differently put, total income of the richest 25 million Americans is equal to total income of almost 2 billion people' (Milanovic, 2002, p. 51). In short, while the median household incomes of a few countries increase, this is not a usual occurrence, and the divide between the world's rich and poor nations is far from disappearing. As a corollary, the perception of the world as being divided into 'haves' and 'have-nots' is persistent and widespread, especially in the developing countries.

The history of climate governance

Poverty, both in its relative form, in comparisons with the developed world, and as absolute poverty or lack of basic capabilities (Sen, 1999) results from global inequality and directly impacts on the ability of developing countries to negotiate with developed countries. A useful indicator for this issue is the average number of delegates sent to climate change negotiations. At the Sixth Conference of the Parties of the UN Convention on Climate Change, the Alliance of Small Island States (AOSIS), for example, was represented by 3.3 delegates per country on average, while the EU countries were represented by 40.6 delegates on average, some of the major EU countries being represented by delegations of 100 persons or more (Dietz, 2009, p. 202; see also Richards, 2001). Roberts and Parks (2006, p. 17) note that delegations from developed countries normally attend these conferences with a 'convoy of lawyers, legal experts, scientists, economists, skilled diplomats, and observers, allowing them to read every document, attend every committee meeting, and painstakingly weigh the pros and cons of proposals'. In contrast, representatives from developing countries typically have to make great efforts to attend meetings and study conference materials in depth. It is often difficult, if not impossible, for them to process information, let alone respond to it in time. This imbalance is aggravated by the fact that climate negotiations are normally divided into committees and sub-committees in order to cover the wide range of issues raised. Since many issues are usually dealt with at the same time, agreements are 'routinely struck in the absence of [the] under-staffed and overstretched' (Roberts and Parks, 2006, p. 16) governments of the least developed countries. This asymmetry is also expressed in the huge public budgets that permit developed countries to put together large delegations, which can negotiate effectively in many places at the same time, while the limited budgets of developing countries make it difficult for them to pay the salaries and accommodation fees for delegation members or the bills for the legal, economic, environmental and social policy advice that these members need in order to follow proceedings and to respond to the demands of their countries.[6] Furthermore, varying abilities to fund research follow from differences in budget and constitute a major factor towards explaining why, to date, climate science has been predominantly 'Northern'. Kandlikar and Sagar (1999) examined the cross-national distribution of authors in International Panel on Climate Change (IPCC) Working Groups in 1995 and found that, out of the 512 authors in Working Group I, 212 were from the US, 61 from the UK, while just 12 authors came from China and India combined.

Hence, international negotiations on climate change do not take place within an 'ideal speech situation' (Habermas, 1990). Though formally characterised by equality among delegations, climate talks are overlaid by the structure of global inequality, stakeholders' material interests and the power asymmetries between states. Trust – a necessary precondition for functioning governance structures – is indeed difficult to establish in a situation where most decisions require formal consensus but where, in practice, the delegations from the developed countries command formidable agenda-setting power, while the delegations from developing countries very often have no alternative but to 'fall back on rhetorical statements, rather than making concrete problem-solving proposals' (Roberts and Parks, 2006, p. 16; see also Gupta, 2000), due to their lack of economic, legal, scientific and political resources. Given the power asymmetry inherent in the existing governance structure, it could be hypothesised that both the outcomes of international negotiations and the means and mechanisms selected to deal with climate change will largely reflect the interests of those parties that are already privileged within the existing structure of the global division of labour: the developed countries and the multinational corporations, which usually have their headquarters in one of these countries.[7] In order to verify this hypothesis, it is useful to reconstruct the historical emergence of the governance edifice of climate policies, with special emphasis on the different capabilities of the negotiating actors to achieve their interests in the process.[8]

The first World Climate Conference, in Geneva in 1979, sponsored by the World Metereological Organisation (WMO), was essentially a scientific conference covering a wide range of disciplines. During the conference, four working groups were set up for the assessment of climate data, the identification of climate topics, the conduct of impact studies and research on climate variability and change. Furthermore, it was agreed to establish the World Climate Research Programme (WCPR), which was to be coordinated by the WMO and the United Nations Environmental Programme (UNEP). The conference was a crucial step towards the creation of the IPCC by WMO and UNEP in 1988. The Second Climate Conference was held in 1990, again in Geneva. In the light of the IPCC's first assessment report, delegates issued a strong statement that highlighted the risk of climate change. Debates and developments at the conference led to the establishment of the United Nations Framework Convention on Climate Change (UNFCCC), to which the Kyoto Protocol belongs. This framework was signed by 154 states at the first United Nations Conference for Environment and Development (UNCED) in Rio de Janeiro in 1992, informally known as the Earth Summit. The objective of the treaty was the stabilisation of greenhouse gas concentrations in the atmosphere at a level that would prevent dangerous anthropogenic interference with the climate system. This objective came into force in 1994, after ratification by over 50 signatory countries. The signatory countries have met annually since – in Conferences of the Parties (COP), beginning with COP1 in Berlin in 1995. The Berlin Summit resulted in a UN ministerial declaration that determined the path of future negotiations: after a two-year analytical and assessment phase, a comprehensive menu of actions that would enable countries to address climate change were to be decided. During the first stage of implementation, following the principle of 'common but differentiated responsibilities', developing countries were exempted from introducing measures to reduce carbon emissions. In 1995, the IPCC also published its second assessment report, suggesting 'a discernable [sic] human influence on global climate change' (1995, p. 22). This and other IPCC findings found their way into the proceedings of the COP2 meeting in Geneva in 1996. In relation to the measures that countries were advised to take, the summit rejected 'harmonised policies' in favour of 'flexibility' and called for 'legally binding mid-term targets' that were to be agreed on at the following summit (COP3) in Kyoto in 1997.

On that occasion, approximately 10,000 delegates – including 170 government delegations, several hundred non-governmental organisations (NGOs) and a multitude of international organisations – took part. Binding decisions, however, were brokered and taken by the participating government delegations only. The 'Kyoto Protocol' finally adopted envisaged the following measures. First, an overall reduction in greenhouse gases of at least 5.2 per cent was projected for the period 2008–2012, which was defined as the first emissions budget period. Second, most industrialised nations (defined as Annex A countries) and some central European economies in transition (Annex B countries) agreed to legally binding reductions in greenhouse gas emissions to an average of 6–8 per cent below 1990 levels for the same period. Third, the so-called 'flexible mechanisms' – carbon trading, Joint Implementation, the Clean Development Mechanism – were introduced, allowing industrialised countries to fund emissions reduction activities in other locations, including in developing countries (on this, see Chapter 13).

Different authors evaluate the outcomes of Kyoto in different ways. Some highlight the fact that the Kyoto Protocol marks a milestone in the international governance of environmental policy and of climate change in particular. For the first time in history, legally binding environmental goals were agreed upon and a foundation for further international efforts to reduce carbon emissions was laid (Brunnengräber et al., 2008, p. 91). Yet most observers point to the necessary CO_2 reduction targets identified by the IPCC (see the Introduction) and agree with the majority of scientists, who regard the benchmark reduction of 5.2 percent as not ambitious enough to avert life-threatening climate change. Nor did most developing countries view the UNFCCC and the Kyoto Protocol as a 'great victory'. Najam et al. (2003, p. 222), for example, argue that, 'in deciding to set first period emission targets as a percentage of 1990 emissions rather than as an allowance of emissions per capita, the Kyoto Protocol has set an allocation precedent which benefits those with high current emissions rather than those whose current emissions are low'. Developing countries, most of which have relatively low per capita emissions, were saddled with much lower emission allowances than their counterparts in the developed world.

Post-Kyoto negotiations have remained difficult and have focused, above all, on the implementation of individual aspects of the Kyoto Protocol. Parties at COP4 in Buenos Aires in 1998 adopted a two-year action plan to advance efforts and to devise mechanisms for implementing the Kyoto Protocol. When COP5 in Bonn in 1999 did not lead to major progress, a breakthrough was achieved at COP6 in The Hague, in 2000, regarding the further particulars of issues such as emissions trading, allowances for carbon sinks in forests and agricultural lands, consequences for countries' non-compliance in meeting their emission reduction targets and financial assistance for developing countries to deal with adverse effects of climate change. Yet, again, the gaps between the positions of the different parties could not be breached, and talks were resumed in Bonn in 2001. Between The Hague and Bonn, the IPCC published its third assessment report, which essentially confirmed the previous result, stating that there were anthropogenic causes of climate change. During the same period, the new Bush Administration in the US announced its retreat from the Kyoto Process and refused to ratify the Protocol, pointing to negative effects for the US economy from the implementation of the agreement. Other states used this as an opportunity to distance themselves from the process too, thereby endangering the formation of an international climate regime (Brunnengräber et al., 2008, p. 92). Expectations for further progress were correspondingly low, and it was somewhat surprising that agreement on several controversial issues was reached at the summit in Bonn. Not only were the details of the flexible mechanisms defined, but also allowances related to carbon sinks were specified for a range of activities that store carbon from the atmosphere, including forest and cropland management and re-vegetation. The COP7 meeting in Marrakech further defined the operational rules for international emissions, trading among parties to the Protocol and for the clean development mechanism (CDM) and joint implementation (JI) (see Chapter 13). However, a concrete limitation of the CO_2 reduction certificates allowed to each country via the CDM and JI in order to meet CO_2 emission reduction benchmarks was not defined (UNFCCC, 2001). Thus, a back door remained open for industrialised countries to 'outsource' CO_2 reduction rather than to initiate structural changes in their domestic economies (see Chapter 13). COP8 in New Delhi (2002) and COP9 in Milan (2003) merely led to minor amendments in relation to the implementation of the treaty. COP10 in Buenos Aires in 2004 was accompanied by the imminent ratification of the Kyoto Protocol after the Russian Duma had ratified it.

Once the Kyoto Protocol was finally in place, the period after 2012 – when the national commitments of the first round were due to end – came into focus. But, after Buenos Aires, international climate negotiations became even more complex. In addition to the annual meetings of the signatory states of the UNFCCC – the COP summits – meetings of the parties of the Kyoto Protocol (MOPs) began to take place. Thus, COP11 in Montreal in 2005 was also the first meeting of the parties (MOP1) of the Kyoto Protocol since their initial meeting in Kyoto in 1997. Further complexity in the multi-level governance system of climate policy was created by the US government, which reacted to the ratification of the Kyoto Protocol with the foundation of the Asia Pacific Partnership on Clean Development and Climate. Brunnengräber et al. (2008, p. 95) interpret the launching of this alliance as an attempt to move the attention away from international climate policy towards bi- and multinational partnerships, which are characterised by voluntary – that is, non-legally binding – measures on the part of the private sector. According to these authors, the US initiative aimed at undermining the commitments of the Kyoto Protocol and at convincing Asian countries of US proposals. At the same time, partly due to the US's rejection of and self-exclusion from the Kyoto Process, climate policy became an increasingly more important issue at G8 summits, especially at Gleneagles in

Scotland in 2005 and at Heiligendamm in Germany in 2007. Given the increasing complexity of governance structures and the diversity of the interests of the major players involved, it is not entirely surprising that no agreement was reached on a post-2012 framework for the successor of the Kyoto Protocol during the most recent COP and MOP summits in Nairobi (2006), Bali (2007), Poznań (2008), Copenhagen (2009) and Cancun (2010).

Expectations were especially high prior to the Copenhagen 2009 summit. Since the US had a new President – Barack Obama – who had promised to re-evaluate internationally coordinated political action and in particular the UN as the main decision-making institution at international level, there was widespread hope that an ambitious global climate agreement for the post-2012 period could be reached. Ministers and officials from 192 countries travelled to Copenhagen in December 2009. A record number of participants from a large number of civil society organisations likewise took part in the negotiations. Thousands of grassroots organisations and some 50,000 ecologically committed individuals gathered at the alternative Klimaforum09 and engaged in over 200 panel discussions, information meetings and other debates; a demonstration with 100,000 participants was held and many other formal and informal exchanges on the possibilities of low-carbon lifestyles and transition towards a non-fossil economy took place (Foreningen Civilsamfundets Klimaforum, 2010). Yet all attempts to keep up the pressure on delegations at the UN conference to broker an acceptable deal appeared to have been in vain when the summit ended in almost total failure. Far from the summit achieving a workable and legally binding agreement for the post-Kyoto period, its only outcome was a 13-paragraph 'political accord' negotiated by 25 parties, including the US and China. This accord was merely 'noted' by the COP, due to a lack of consensus among delegations. It contained a collective commitment from developed countries to new and additional resources and to investment through international institutions approximating $30 billion for the period 2010–2012. Due to the non-binding nature of the Copenhagen Accord, expectations were correspondingly reduced for COP16 and MOP6 in Cancun, Mexico in 2010. The outcome of the summit was an agreement, though not a binding treaty. Again, the parties recognised that the warming of the climate system is unequivocal and that climate change represents an urgent and potentially irreversible threat to human societies, which must be urgently addressed. The agreement further reiterated the goal of reducing global greenhouse gas emissions so as to keep the increase in global average temperature within the limit of 2°C above pre-industrial levels. While the parties realised that addressing climate change requires a paradigm shift towards building a low-carbon society, there was no agreement on how to extend the Kyoto Protocol. The agreement includes the set-up of a 'Green Climate Fund' worth $100 billion a year by 2020 to assist poorer countries in financing emission reductions and adaptation. However, it was not specified how this annual sum would be raised. Unresolved issues were referred to working groups, which were due to report at forthcoming COP and MOP summits, beginning with COP17 and MOP7 in 2011 in Cape Town, South Africa.

The main actors

The modest results of international climate negotiations to date must be understood within the context of the increasingly complex list of stakeholders, which includes, among others, national governments; coalitions and various alliances of states (EU, G8, G20, e.g.); advisory bodies such as the IPCC and the Climate Change Secretary in Bonn; institutions and interest groups with observer status (several UN organisations, the OECD, the International Energy Agency (IEA)); a multitude of more or less connected NGOs; and several employers' and employees' organisations. The remainder of this chapter is, therefore, dedicated to an understanding of the diverse interests of these stakeholders.[9] National governments are of special importance since no other actor is authorised to sign international treaties. However, following Mayntz (2009), the main interest of national governments is to avoid international agreements that compromise the competitive interests of powerful private actors within their own territory. This partially explains why compromises on the lowest common denominator are the rule in the multi-level governance system of climate policy. For a deeper analysis of interest conflicts between national governments, it is useful to build upon the UNFCCC's distinction between three groups of signatory states: Annex A countries comprise both industrialised and threshold or newly industrialised countries, while the industrialised countries alone count as Annex B countries: those that were OECD members in 1992. Threshold countries include Russia, the Baltic States and most Eastern and South Eastern European countries. The developing countries are subsumed under the category Non-Annex A countries. The expectation expressed in the UNFCCC was that Annex B countries, given their historical responsibility, would make the financial means available for adaptation and mitigation to climate change – that is, not only for themselves, but also for developing countries.

Yet the UNFCCC's division into three country groups is still too broad to reflect the actual interests of and power relations between countries and country groups. State coalitions in climate negotiations often follow different lines, so that there are both 'foot-dragging' and 'progressive' forces in relation to climate protection in all three UNFCCC categories. For a further distinction between Annex B countries, Brunnengräber et al. (2008, pp. 99–101) divide the US, Japan, Canada, Australia, Switzerland, Norway and New Zealand into one group and the EU countries into a second. The former group was, for a relatively long period of time, united in its refusal of any legally binding commitment in relation to climate change protection. Yet the group was divided, when the US left the Kyoto Process, while Japan, for its part, continued to participate. A further split arose when Australia ratified the Kyoto Protocol under its newly elected Labour government in 2007. The EU, in contrast, has enjoyed a positive reputation in climate negotiations for some time. With its proposal of an overall 20 per cent reduction of carbon emissions by 2020 and with the announcement that it would increase this target still further if other industrialised countries would make similar efforts, the EU, as a whole, is indeed comparatively ambitious in its CO_2 emission reduction plans. Yet the label 'progressive force' disguises the fact that EU member states have always had differing interests and policy proposals. Denmark, Sweden, the Netherlands and, to a lesser extent, Germany and the UK tend to push for relatively ambitious benchmarks in relation to climate protection, while the Southern European countries stress on their right to economic catching-up development. The new Eastern European EU member states form yet another group, as they belong in the threshold or Annex A countries. The EU directive on the implementation of the Kyoto benchmarks reflects the different situations of EU member states by following the principle of 'burden-sharing' and by defining different greenhouse gas reduction targets for different countries. The EU's capacity of speaking with one voice in climate policy is further complicated by the fact that some EU countries are also members of the G8, which recently demonstrated increased interest in climate policy, albeit mainly in conjunction with energy supply protection policies (Brunnengräber et al., 2008, p. 100).

The developing countries have some common interests in climate negotiations when compared to developed countries, but they are also divided on several specific issues. The common interests of the developing countries are expressed by the G77 and its demand to link climate policy to structural reforms of the global economy. The idea of the 'right to development', for instance, was combined with the demand for an individual greenhouse gas emissions account that would apply to each inhabitant of the globe; the demand for debt relief was issued in conjunction with the admission of historical guilt and responsibility for climate change by the North; and the recognition of the atmosphere as a common good was linked to consideration of the specific needs of local communities in developing countries with respect to their adaption to climate change (Dietz, 2009, p. 196). However, Mayntz's pessimistic assessment that short-sighted national competitive interests normally outweigh long-term 'global' concerns when international negotiations are being held also seems to apply to the developing world, whose divisions follow largely the logic of specific export interests. Since the economic development of countries that export raw materials, and especially fossil energy carriers, is undermined to a much greater extent by climate protection measures than the economies of countries that do not (Brunnengräber et al., 2008, p. 98), it is no coincidence that the OPEC countries form an almost homogenous block in climate negotiations. United in their interest to sell oil in the world market, they are viewed by most observers as foot-draggers in relation to climate protection. The OPEC countries have formed coalitions with the US on many occasions. A further important division among the developing world countries is that between threshold countries, whose development has proceeded rapidly and whose carbon emissions have simultaneously increased enormously – China, India, Brazil and South Africa – and the least developed countries, whose socio-economic development has stagnated in recent years and who have contributed next to nothing to climate change due to very low-carbon emissions. Finally, there are those countries with a high risk of drought and desertification, and those that are existentially threatened by sea-level rise and do not have access to the means necessary to avert life-threatening climate change. The AOSIS group of states, in particular, tends to favour the most ambitious climate goals, yet it lacks bargaining power at climate summits. This is why these states have no alternative but to make moral appeals to the world public. Most famously perhaps, on the occasion of the UK premiere of the film *The Age of Stupid* by Franny Armstrong on 15 March 2009, Mohamed Nasheed, President of the Maldives, pointed out that his home country is under threat of sinking due to the prospect of rising sea levels. In the same speech, he announced that the Maldives would strive to become the world's first carbon-neutral country.

Apart from national governments, a range of further actors partake and influence climate negotiations. First and foremost the IPCC, whose task is to provide scientific information on climate change for policy-makers (see the Introduction). However, due to the fact that the IPCC's assessment reports are the most important scientific foundation upon which COP decisions are made, they also wield a significant amount of informal political power (Brunnengräber et al., 2008, p. 103). Second, the annual cycle of COP and MOP summits could not exist without the Climate Change Secretariat, which has its seat in Bonn and currently employs more than 200 people. The tasks of the Secretariat include preparation of the annual meetings of the signatory countries, monitoring of whether the agreed climate policy targets have been met, consultancy in relation to policy implementation and coordination with other UN bodies. Third, there are several organisations that the climate secretariat in Bonn recognises as having observer status at COP and MOP summits. These include UN organisations such as the UNEP and the United Nations Development Programme (UNDP), and further international organisations such as the WMO, the OECD and the IEA. Many NGOs, which historically played an important role in raising awareness of the issue of climate change, likewise have observer status. By their very nature, NGOs are diverse. About 400 NGOs coordinate their action in the Climate Action Network (CAN), supporting the Kyoto Process and providing consultancy to policymakers, in particular with respect to the flexible mechanisms, while others are more critical of the mainstream governance of climate change. According to Dietz (2009, p. 203), the latter group of NGOs is often denied access to national delegations. Critics of the expression 'NGO' argue that it has become so broad that it is sometimes applied even to the traditional lobby organisations of the energy or car industries, which also partake in climate talks. At COP13 in Bali, for example, Dietz (ibid.) counted 350 delegations representing major corporations. These did not attempt to block negotiations, but demanded instead a continuation of the Kyoto Process, and especially of the CDM. Different interests between lobbyists are also reflected in the negotiations (Brunnengräber et al., 2008, p. 106). The insurance industry is especially affected by disasters associated with climate change and with corresponding financial claims from customers while established multinational energy concerns tend to be biased towards continued use of fossil fuels and, therefore, advocate traditional energy policies. An analysis of the regional composition of the 860 organisations recognised as having observer status in climate summits by the Climate Secretary in 2008 (Dietz, 2009) found that only 160 were based in developing countries, and of these more than half came from threshold countries such as Argentina, Brazil, India and South

Africa. Dietz's research therefore confirms this chapter's main argument, that the mere existence of global policy networks and governance structures does not guarantee the inclusion and consideration of the interests of the weakest social groups and of the least developed countries. The structures of global social inequality remain omnipresent in the current international governance network on climate change.

13
Theory and Practice of Carbon Emission Trading: The Case of the EU ETS

This chapter deals with the procedural aspects by which the – rather modest – goals of the Kyoto protocol are to be achieved. Generally, the choice of one certain policy path from a range of options is not independent of wider political and societal debates and dominant ideologies. The spread of neoliberal ideas and solutions from a tiny circle of economists in the Mont Pelerin Society to a largely undisputed worldview is a good example (see Chapter 8). Trading systems were integrated worldwide, property rights systems were restructured and socio-economic regulation facilitated transnational corporate activity, weakening the regulatory power of national corporate actors. Developed in well-funded think tanks and business schools across the world, the neoliberal perspective proposed the transformation of a range of public goods into privately held commodities. These served as investment opportunities for the liquid capital that was accumulated at an unprecedented extent in finance-driven capitalism. At the height of this transition, neoliberalism indeed took the character of a new *pensée unique* (Bourdieu) and was not only applied to economic regulation, but also to other fields, including housing and pension policies. Environmental policies, and the Kyoto procedures in particular, constitute no exception to this rule, as they followed arguments first developed by Ronald Coase in the 1960s.[1] Coase suggested that environmental protection should not be safeguarded via taxes and state regulation. Instead, it should be left to private actors in markets that had yet to be established.

However, over a long period, the environment did not figure prominently in mainstream economic theories. In theories on the factors of production, the environment was usually subsumed under the factor 'earth'. During the twentieth century and as environmental problems

became difficult to ignore, economists developed a new field of specialisation – environmental economics. The basic assumption in neoclassical
economics is that markets are a good way of allocating goods and that
they tend towards general equilibrium, but that there are cases in which
market regulation leads to non-intended results. Market outcomes are
then distorted by exogenous factors or 'externalities', which arise 'whenever one individual's action affects the utility of another individual'
(Cowen, 1988, p. 2). Environmental problems such as contaminated
air are seen as matters of negative external effects, since they are not
expressed in terms of market prices in the first place. Such negative
impacts for third persons emerge as a result of the economic actions
of others without any possibility of holding the perpetrators to account.
This, for example, is the case when a power-generation company cannot be made accountable for the air pollution it causes for a legal or
economic reason. In order to solve this problem, various economists
have suggested 'internalising' external effects, that is, assigning them to
their initiator following the cost-by-cause principle. However, the cause
and the effect of an ecologically harmful economic activity are often
difficult to determine. It is, for example, almost impossible to identify a
single perpetrator responsible for the extinction of a species (Ptak, 2008,
p. 37). Due to this difficulty and in order to avoid complicated legal proceedings, neoclassical economists deem necessary a monetary valuation
of external effects, so that they can be expressed in commodity prices.

Three methods in environmental regulation

The 'internalisation' of external effects can proceed, in principle, using
three ways or methods.[2] The first method concerns directly regulative measures, where the government establishes restrictions on how
much pollution a company can emit. Exceeding the allowed values
of a particular kind of pollution identified by government authorities
leads to penalties ranging from fees to the closure of the emitting
industrial unit, and so companies have an incentive to review their
production methods. Costs such as those for potential fees, for the
adaptation or outsourcing of production methods become part of companies' financial calculations; they are thus 'internalised'. The second
method is the imposition of taxes on the producers of negative external effects. This principle is known as Pigouvian tax, after Arthur Cecil
Pigou (1932). Since negative externalities arising from certain economic
activities lead to damages for third parties and to costs for the general public that are not already covered by the private costs of the

company, the activity in question is taxed in order to 'correct' the market outcome, so that efficiency is achieved. Following Pigou, governments should use the tax income they raise to compensate for damages caused and for the financing of measures taken against the causes of the damage.[3] Pigou's critics pointed out that his approach was not easily applicable in practice since economies are normally not stable and static but tend to grow (Nell et al., 2008). During this process, the material composition of commodities and services as well as the types of economic activities change constantly, so that policymakers lack exact knowledge of the gaps between private and external costs. This complicates the implementation of taxes and the precise definition of both 'the object' and 'rate of taxation'.[4] Neoclassical environmental economists further criticise Pigouvian taxation and direct regulative measures because the government, and not the market, is in control of both procedures and outcomes. This is seen as restricting innovation and as being inefficient too. The third method of 'internalising' external effects, therefore, gives priority to the market by creating tradable rights or pollution allowances. Not only does the neoclassical tradition, and the Chicago School in particular, regard this as more in line with market economies and property rights (Demsetz, 1969), but policymakers are also increasingly attracted to market solutions, since these appear to require less political intervention and companies lose less profit than under Pigouvian taxation regimes. Indeed, if allowances are allocated to companies for free, polluters can even make a profit by selling them.

Pollution-trading or emission-trading schemes are essentially an application of the ideas of Ronald Coase. Both Pigou and Coase viewed pollution such as disturbance through noise, unpleasant smells or light, and the use of air space, as examples of negative market externalities. Coase (1960, p. 15), deviating from Pigou, however, suggested constructing specific property rights in order to identify and separate the affecting and affected parties in relation to ecological damages and to calculate these economic costs. Presupposing a 'perfect market' with no transaction costs,[5] Coase argued that the introduction of private trading of allowances would enable affected parties to decide for themselves if, how and to what extent they should restrict environmentally harmful activities. Instead of a costly and cumbersome bureaucracy that taxes negative external market effects, a market would be created for the trading of rights and certificates. As in Pigou's tax regime, perpetrators of environmentally harmful actions would be disciplined but, according to Coase and his followers, in much cheaper and more effective ways.[6] External costs that were previously met by the public would be internalised, and so companies would be confronted with the real costs of their actions and would change their production methods accordingly. Pollution rights were to be made transferable in order to make the mechanisms of demand and supply function properly. Hence, those companies that could not easily function without creating pollution would acquire the right to do so from others, for whom emissions reduction was easier to achieve. For Coase, another advantage of market solutions over tax solutions was that policymakers no longer needed to decide which economic activities were to be taxed and under what conditions. The market would take care of this through the demand and supply of emissions certificates. Finally, Coase was optimistic about the link between pollution, business costs, innovation and technological progress. Rising costs caused by the increasing scarcity of production factors that were previously free would lead to adjustments in the technological and energy basis of the work process, thereby giving optimal ecological results.

John Harkness Dales was an important successor of Coase and a precursor of carbon emission trading. Like Coase, Dales (1969) opted for the reduction of air pollution by applying market mechanisms, following the parameters of economic scarcity. Yet he deviated from Coase by not leaving the definition of the best overall level of pollution to an imaginary 'perfect market'. Instead he gave this task to the government. Pollution trading became, then, a method of finding the most cost-effective way to reach emission goals that had been politically defined in advance. Once the state had defined the total level of emissions for a sector of the economy and a specified time period and then issued proportionate emissions allowances, these could be traded freely between economic actors. The assumption was that those who faced the highest cost of emissions reduction would be prepared to pay the most for the allowances. Those who had at their disposal comparatively cheap opportunities for emissions reduction would opt for taking advantage of these rather than purchasing permits. In short, emission trading promises that 'whatever level of emission control is politically required can be achieved in the most efficient way, at minimal cost to society. Or the other way round, each dollar spent on emission control produces the highest possible effect on the environment' (Voß, 2007, p. 99; see also Baron and Philibert, 2005). With regard to the ecological effects of the mechanism, it is crucial for government authorities to ensure that allowances are issued in such a way that their number is reduced over time, so that the price for each emission unit rises sufficiently to create an economic incentive to implement ecologically desirable innovation in the work process (Ptak, 2008, p. 39). In order to keep allowances scarce, Dales suggested auctioning them to companies.

Carbon markets

A prominent contemporary supporter of applying market solutions to climate change is Nicholas Stern, who proposes three arguments that prove the superiority of market solutions over carbon taxes.[7] First, Stern argues that taxes would not provide much certainty on the level of CO_2 emissions reductions since estimates of responses to taxes are 'imprecise and many of the effects operate with long lags' (Stern, 2009, p. 102). Second, he argues that taxes are very hard to coordinate internationally, as countries are 'very protective of their tax independence'; and, third, he views electorates as generally 'mistrustful of governments' use of tax revenues' (p. 103). In contrast, the definition of CO_2 emissions reduction targets and carbon trading as the method of meeting these targets would allow for 'greater certainty about the quantity of emissions, because that quantity is directly set by the government' (p. 104). If the price for carbon emission allowances is set in the right way, then consumers and producers would have an incentive to cut emissions, since the price of a commodity would not only embody the 'cost of the raw materials, labour, capital and so on used in its production, but also the cost of the damages from the emissions produced in the consumption or production of the good' (p. 100). In Stern's view, another advantage of the price mechanism is that it achieves greenhouse gas reductions as efficiently as possible. The creation of a single price for emissions would ensure 'that all opportunities for reducing emissions which cost less than the price will be exploited' both within and across countries (Stern, 2009, p. 100). This argument is linked to a third advantage of trading schemes, which Stern sees in 'international private-sector flows of carbon finance from rich to poor countries' (p. 104). Carbon-trading schemes are seen to have an environmentally positive effect via the application of innovative technologies in developing countries. Consequently, while Stern acknowledges that governments should be free to find the policy solutions that fit their countries' situation, he nevertheless advocates a 'strong role for trading schemes' for every country in the world (p. 105).

Stern accords much less attention to the problems of implementing carbon-trading schemes, when these are compared to his celebration of market virtues and their potential impact on the reduction of greenhouse gases. However, in contrast to other free-market environmental economists, he at least mentions two of them. First, he insists that initial allocations of CO_2 emissions allowances must not be too high; otherwise 'reductions in emissions and prices will both be too low'. According to him (p. 108), there is 'no alternative to tough allocations', and "'special pleading" or "gaming" must be resisted'. However, he does not discuss how exactly these circumvention strategies can be avoided other than through auctioning emissions allowances. The second problem, mentioned by Stern, with the practicability of market solutions is the clean development mechanism (CDM) as defined in the Kyoto Protocol. Though he applauds the scheme in general, he criticises the need for companies to prove that their measures to reduce greenhouse gas emissions are more than what would otherwise have been done, which is according to Stern, 'just too cumbersome for administration on a large scale' (p. 110). Yet his overall assessment of the implementation problems of carbon-trading schemes is that these are 'teething troubles', which can be overcome if applied intelligently. With regard to the future, he shows enormous optimism and envisages that a worldwide cap and trade system for carbon emissions – that is, for both the developed and the developing world – will be in place by 2020 (Chapter 16). This optimism is partially due to the fact that the 'economist' Stern fails to discuss the structures of global inequality and of power in international relations as reflected in the international climate talks (Chapter 12). In his publications, as Giddens (2009, p. 201) remarks, it is 'as if the "global deal" will be reached as soon as the nations of the world see reason'. And Giddens' sociological riposte to Stern's demand that 'all must play their part' is to ask 'who is there to implement the "must"?' Stern's faith in carbon markets can be almost unlimited, precisely because he ignores the realities of inequality and domination at global level.

From a regulation theoretical perspective, it appears far from accidental that neoliberalism made its mark upon environmental economics. The view that market solutions are, almost by definition, superior to taxation or state legislation reflects a wider debate in economy and society. Yet the commodification of something that is not scarce, such as air, or that has no use value, such as CO_2,[8] illustrates particular problems. Since the existence of markets for tradable entitlements to emit a certain amount of CO_2 cannot be taken for granted, these markets must be created by an active state and its agencies (Ptak, 2008, p. 38). It is only when private property rights to contaminate the atmosphere exist that specialised platforms for emissions trading can emerge – platforms that are accessible to holders of emissions certificates worldwide. 'Compatible with the established financial market regime and its regulations and routines' (Voß, 2007, p. 103), the Chicago Climate Exchange (CCX) was the first system for the registration, reduction and trade of greenhouse gas emissions: a self-regulated stock exchange managed by its members.[9] Emission certificates are also traded as EU Allowances (EUAs) on stock exchanges in Europe. In Leipzig, for example, currently available EAUs are sold and purchased at their current prices, and there is also a Forward Market for contracting the purchase of EAUs at future dates and at fixed prices (Ptak, 2008, p. 41). Through the implementation of carbon-trading schemes, new business and investment opportunities arose for CO_2 brokers, tradesmen and bankers, including those representing major finance companies and hedge funds. New actors who are becoming involved in the implementation and operation of carbon-trading schemes include market intermediaries, auditing companies, consultants, lawyers and various researchers. Though conflicts between environmental goals and business interests have not completely disappeared, there has certainly been some 'rapprochement', since emissions trading schemes bypassed the 'established regime of command-and-control regulation' and linked up instead with the 'rise of market-oriented regulatory reform as a dominant agenda on the basis of neo-liberal governance concepts' (Voß, 2007, p. 118), ensuring the support required from business circles. Neoclassical economists and mainstream climate change governance theorists do not find it problematic that investors are not primary interested in reducing atmospheric CO_2 concentrations but rather in the financial returns arising from the trading of and speculation with certificates. On the contrary, since the reduction of CO_2 emissions is expected to be a side-product of merely furthering individual profit interests, emissions trading schemes are regarded as a welcome new investment opportunity for financial capital.

Historically, emissions trading schemes were first developed and used in the US. Early examples include the US Environment Protection Agency's scheme, which followed the Clean Air Act Program to improve local air quality of 1974, and the Lead Training Program of 1973, which aimed at reducing the percentage of lead in petrol. The Environment Protection Agency issued allowances of permissible lead content and allocated them to refineries. The latter saved their allowances if they met the reduction benchmark in advance, and so they could use these allowances later (Schreurs, 2008, p. 24). Emissions trading became widespread and accepted as an environmental policy instrument in the US during the 1990s. The first international emissions trading scheme was the Montreal Protocol, which referred to substances that contributed to the decomposition of the ozone layer. It contained time schedules for the reduction and, in some cases, the complete abolishment of several ozone depleting substances, especially chlorofluorocarbon. Different benchmarks were defined for developing and developed countries, the former receiving assistance in monitoring the Montreal Protocol from a Multilateral Fund managed by the UNEP Ozone Secretariat. However, throughout the 1980s and early 1990s, the expansion of emissions trading in international environmental policy was blocked by scepticism and resistance from European policy circles, which were worried that environmental goals would be subordinated to private profit interests. Paradoxically, transnational corporations such as the oil companies BP and Shell became pioneers in the implementation of emissions trading systems at company level. This provided a transnational precedent and functioned as vanguard for this new policy instrument to be imported into EU policy networks (Voß, 2007, pp. 108–9). Towards the end of the 1990s, the Organisation of Economic Cooperation and Development (OECD) started to endorse emissions trading actively and to offer review, evaluation and dissemination services for carbon trading programmes. Voß outlines how developments at different levels of governance resulted in a 'global hype' around emissions trading, which was soon viewed as 'the' environmental policy instrument of the future. In a U-turn of their previous criticism, European policymakers joined in and made 'widespread attempts to become part of the emerging movement' (p. 109). The new policy instrument appeared especially promising since, in the context of the negotiations of the Kyoto treaty, 'foot-dragger' countries that basically objected to any kind of legally binding agreement could now be brought on board. For this to happen, sufficiently 'flexible' mechanisms had to be proposed that guaranteed the most effective, that is, cost-effective ways to meet climate targets (Brunnengräber et al., 2008; Haensgen, 2002; Lohmann, 2006).[10]

On the basis of the carbon emissions reduction targets specified in the Kyoto Protocol, each Annex B country that emits more CO_2 than assigned and, therefore, has to reduce emissions receives 'assigned amount units' or allowances. The Protocol defines two instruments, which both follow the principle that emissions reduction should be carried out at locations where this reduction is the most cost-effective. The two instruments are the 'joint implementation' (JI), whereby such measures are implemented in other Annex B states, and the 'clean development mechanism' (CDM), whereby projects are carried out in cooperation between developed and developing countries (Ebert, 2010). According to neoclassical environmental economics, this type of cooperation translates into a win–win situation for both parties involved. Companies from Annex B countries are presented with three options in any combination – 'cutting their own emissions, trading allowances with each other, or buying credits from abroad' (Lohmann, 2006, p. 48) – whereby they join or entirely finance a project in a developing country and receive 'credits' in correspondence to the amount of allowances this project saves in return. Developing countries, for their part, are assisted by the developed countries, through technology transfer, towards a more sustainable economy. Though the Protocol mentions that emissions trading should be supplementary to national measures and many developing countries lobbied for a 50 per cent ceiling for CO_2 emissions reductions within national borders, the Protocol does not quantitatively specify this ceiling.

The EU experience

The EU experience is especially relevant when evaluating the efficiency of carbon-trading schemes in relation to climate change mitigation since the EU has the largest multi-country, multi-sector carbon emissions trading scheme worldwide: the EU Emissions Trading System (EU ETS). This has been in operation since January 2005 via directive 2003/87/EC. The EU ETS established an allowances market of over two billion tonnes of carbon emissions from some 11,000 economic units or 'installations' (Table 13.1). Taken together, these were responsible for about 40 per cent of CO_2 emissions in the energy sector, mineral oil refineries, coke ovens and the cement, glass, ceramic, steel, paper and cellulose industries. However, this initial limitation of the scheme to

Table 13.1 Verified emissions, average allocations and number of installations in EU ETS (EU 27 minus Romania, Bulgaria and Malta) 2005–2009

Year	Verified emissions (tonnes of C02)	Average allocations (tonnes of CO_2)	Number of installations
2005	2,012,043,453		10,282
2006	2,033,636,557		10,605
2007	2,049,927,884	2,151,926,173	11,186
2008	2,119,602,968	1,957,060,701	12,622
2009	1,873,225,616	1,967,387,261	–

Source: EU Commission (2008a and 2010)

industrial and energy companies caused these industries to claim they were disadvantaged in comparison to their global competitors. Unlike most carbon-trading theorists, who opt for the auctioning of certificates, the EU commission reacted to this lobbying by issuing about 95 per cent of all allowances during the first trading period (2005–2007) and some 90 per cent during the second (2008–2012), all free of charge. The sanctions for installations that emit CO_2 without having any corresponding allowance were determined at 40 euro per tonne during the first phase and at 100 euro per tonne during the second phase. The EU further agreed on the so-called 'burden sharing' principle; here greenhouse gas reduction benchmarks are unequally distributed across member states. Germany and Denmark, for example, agreed to reduce CO_2 emissions that were at an above-average level while Greece and Portugal were allowed to continue to increase their CO_2 emissions. The EU directive was linked to the Kyoto Protocol's 'flexible instruments' – JI and CDM. This made it possible to invest in developed and developing countries outside the EU in order to acquire emission credits for domestic production.[11]

Procedurally, the EU ETS is an iterative process that resembles other areas of European governance where 'soft' and 'open' methods of coordination rather than 'hard' regulation are applied.[12] Member states draw up National Allocation Plans (NAPs), which must then be confirmed by the EU Commission. In order to implement NAPs, member states allocate emission allowances to the relevant CO_2 emitting companies, each operating from its own state territory. Each member state establishes a national registry, to which each installation is required to submit its emissions data. National registries are linked to the Community Independent Transaction Log, which integrates all national systems under a European umbrella, issues allowances and registers accounts for each installation. In cases where allowances exceed the allocated amounts, companies have to acquire additional allowances. Where companies improve their energy balance, they are entitled to benefit from the sale of their superfluous certificates. A major problem in the initial phase of the EU ETS was that most countries had issued too many certificates to businesses (Brouns and Witt, 2008, p. 68), with the result that prices for emissions certificates collapsed. The price for the emission of one tonne of CO_2 fell by two-thirds, from 30 euro in April 2005 to 10 euro in May 2006 (Schreurs, 2008, p. 29). Due to the almost unlimited availability and lack of scarcity of carbon emissions allowances, their trading price hit rock bottom at euro 0.3 in late 2007. With prices for emitting CO_2 approximating zero, it is not possible to speak of an 'incentive system' to implement measures that are likely to abate climate change. It is not surprising that CO_2 emissions from EU ETS installations did not go down but increased during the first EU ETS trading period between 2005 and 2007 (Table 13.1) – namely, from 2,012 to nearly 2,050 billion tonnes. Critics point to the fact that the scheme not only did not deliver progress on CO_2 emissions, but it also became a source for windfall profits for certain industries (Ptak, 2008, pp. 44–5). This was due to the fact that many of the major energy corporations simply transferred the price of allowances, which had been allocated to them free of charge, onto their customers' energy bills. These extra profits appeared perfectly legitimate from a microeconomic perspective, since emission allowances had been allocated a price that became part of the going concern value, just like labour or material costs. Like a baker who has received his or her flour as a gift, but nevertheless sells his or her bread at the going rate, thereby realising extra profit, energy companies happily cashed in the windfall arising from the free allocation of CO_2 emission allowances (Brouns and Witt, 2008, p. 75). In addition, the over-generous allocation of certificates to companies in most of the EU member states allowed many companies to sell on superfluous CO_2 emission allowances on the international carbon markets, which resulted in further windfalls.

During the second trading period (2008–2012), the EU commission modified the scheme in marginal ways, which were unlikely to change the situation fundamentally. Essentially, the percentage of auctioned allowances was raised from five to ten. Ninety per cent of emissions allowances were still allocated free of charge, ensuring ongoing windfalls for energy and other CO_2-emitting companies. Due to the over-allocation of permits, CDM and JI projects were basically not implemented during the first trading phase. The EU Commission had hoped that this would change during the second phase due to the reduction in allowances. Three non-EU members, Norway, Iceland and Lichtenstein, then joined the scheme. Crucial issues, such as the consideration of aviation and greenhouse gases other than CO_2, were postponed until the post-2012 period. The National Allocation Plans of ten member states envisaged an average cut of nearly 7 per cent below the 2005 CO_2 emission levels (EU Commission, 2006). However, due to the generously planned increase in offset credits from JI and CDM projects,[13] it was possible to meet the Phase II cap without any need to reduce CO_2 emissions in the EU space (Committee on Climate Change, 2008). The carbon price recovered a little from its 2007 low point and stood at 13 euro per tonne of CO_2 in the first half of 2009. However, this was still considerably lower than in 2005, when the EU ETS started.

Projections indicate that, like Phase I, Phase II will experience a surplus in allowances (Grubb et al., 2009, p. 12), meaning next to no incentive for installations to modify production processes that would reduce CO_2 emissions in Europe. Consequently, 'Climate Action' EU Commissioner Connie Hedegaard does not attribute the drop in the CO_2 emissions made by EU ETS installations in the years 2008 to 2009 (Table 13.1) to the existence of the scheme, but rather to the severe financial and economic crisis, thanks to which it 'suddenly became easier to reduce emissions' (EU Commission, 2010). The recent reduction in CO_2 emissions is due to the contraction in the general scale of production, and not to innovations in the production process. Hedegaard critically notes that 'European business did not invest nearly as much as planned in innovation, which could harm our future ability to compete on promising markets' (EU Commission, 2010). Future innovations in sustainable production forms are even less likely, since the EU ETS allows companies to 'bank' or save emissions certificates, which are not needed during the crisis and can be used later. In the absence of any innovation in the technological and energy basis of production processes, CO_2 emissions in the EU may well rise again after the crisis – as they did during the period of 2007 to 2008. With regard to the third period of future trading (2013–2020), the EU Commission (2008b) has proposed a number of changes, including the centralised allocation of certificates by an EU authority instead of national allocation plans, a greater percentage of auctioned permits at the expense of freely allocated ones,[14] unlimited banking of carbon emissions allowances over trading periods, the inclusion of the aviation industry and of other greenhouse gases such as nitrous oxide and perfluorocarbons. The proposed cap for an overall reduction of greenhouse gases for EU ETS installations for 2020 is planned to be at minus 21 per cent compared to 2005 emissions. However, it must be reiterated that these changes are still at the draft stage, and it is uncertain whether they become effective from January 2013 onwards. Especially (but not only) in relation to raising the percentage of auctioned certificates and moving the issuing process of certificates from the national to the Europan level, massive lobbying by industrial circles, which has accompanied EU ETS from the beginning (Giddens, 2009, p. 198), will probably water down the Commission's proposal.

14
The Flaws of Free-Market Solutions for Climate Change Prevention and Their Homology to Finance-Driven Capitalism

After the previous discussion of the neoclassical proponents of carbon trading, this final chapter is dedicated to the problematic and counter-productive aspects of this trading. The experience generated to date by available empirical data on actual carbon-trading schemes has already questioned the contribution of these schemes to climate change mitigation. Furthermore, we have noted some of the flaws in carbon trading using the EU ETS – for instance, the over-allocation of certificates and windfall profits for energy companies (see Chapter 13). Against this backdrop, some observers have even dismissed, ironically, the Kyoto Protocol's 'flexible mechanisms' as 'hot air', while apologists argue that both Kyoto and the EU ETS are to be understood as 'learning processes' on the road towards 'more universal and rigorous formulae' (Giddens, 2009, p. 189). We will now discuss whether the flaws and problems of carbon trading systems should be viewed as the 'teething troubles' of an emerging policy instrument or whether they are endemic and likely to persist. I will comment on issues such as the over-allocation of certificates and windfalls for greenhouse gas emitting companies, self-reporting of CO_2 emissions, carbon price volatility, offsetting or CO_2 reduction duties, the problem of 'additionality', geographic distribution, size and content of clean development mechanism (CDM) projects, the bureaucracy and the cost element of carbon-trading schemes and their possibly innovative effect. I will then embed this critique into the wider theoretical framework employed in this book.

Over-allocation of certificates

So far the EU ETS granted more allowances than were needed to cover CO_2 emissions. Even though this maladjustment has been reduced somewhat because of the tightening of the carbon emissions cap during the second trading period, the problem continues to exist. To find out why this is the case, we need not only look at the technical problems arising from the implementation of a new policy instrument, but also consider the competitive position of the EU in the wider economy and the power asymmetries within the EU governance structure. At least initially, the introduction of a regional emission trading system is a disadvantage for European companies in comparison with their global competitors since presently CO_2 reduction benchmarks do not exist in other parts of the world (Ptak, 2008, p. 47). European companies will, therefore, be interested in and lobby for any reduction in their duties to emit less CO_2.[1] The governments of EU member states are in a similar position vis-à-vis the EU commission since, even if they are willing to change, they are nevertheless required to act in ways that benefit the competitiveness of their national locations. It is in their own interest to negotiate a high quota of allowances for their territory as compared to other member states. Fisahn (2008, pp. 60–1), therefore, makes an important point by arguing that the amounts of allocated allowances per EU member state depend not only on the actual CO_2 emissions in these countries, but also on the relative bargaining power of each national government at EU level.

Self-reporting of CO_2 emissions

The method used so far to calculate national emissions, which is at the same time the basis for defining reduction benchmarks for member states, is counterproductive for ambitious CO_2 emissions reduction goals (Lohmann, 2006). This calculation was based upon the status quo of CO_2 emissions in 2005, which, in turn, was defined within the context of self-assessed reports by the operators of industrial units. If one assumes microeconomic rationality and the generation of profit as the main motivation of economic action, then these operators have an intrinsic interest in exaggerating their actual emissions in their reports in order to receive a large share of national CO_2 emission certificates. Initially, exaggerated self-reporting on CO_2 emissions has the welcome knock-on effect, from this point of departure, that any CO_2 emissions 'reduction' targets are easily met, that is, without the need for expensive innovation in the production lines. Yet the opposite case is also conceivable: an operator of an industrial unit producing more emissions than he or she can account for in terms of emission allowances. In this case, the operator will wish to estimate and report low emissions. Again following microeconomic reasoning, the cost for purchasing additional allowances or for upgrading an industrial unit to the prescribed environmental standards will be balanced against the amount of money the operator is likely to pay in terms of sanctions or fines for exceeding allocated emission amounts.[2] Owning too few certificates could be an incentive to incorrect self-reporting, and thus to the undermining of CO_2 reduction commitments by camouflaging real emissions. It is indeed not surprising that 'pandemic cheating' has been highlighted as the Achilles' heel of cap-and-trade approaches in the Human Development Report of 2007/2008 (UNDP, 2007, p. 126).

Carbon price volatility

The price of emissions allowances in the EU ETS collapsed in 2006 and 2007 (Ptak, 2008, pp. 44–5). After a short recovery, prices in the European carbon markets started to decline again from late 2008 against the background of lower oil and energy prices and the deteriorating economic outlook. Such price volatility is not accidental, but typical of energy commodities like crude oil and coal, whose prices are determined at world market level and, therefore, depend on a range of largely unpredictable factors (including major regional crises) that are normally beyond the reach and scale of regional regulation. For example, demand for carbon allowances fell sharply in late 2008 and early 2009 as the recession reduced the economic output, a situation that resulted in much lower emissions and consequently in a lower need for emissions certificates than expected by the EU Commission. The result was what the World Bank (2009) called 'cashing in on carbon during the credit crunch'. A mayor sell-off of European Union allowances (EUAs) started in September 2008, when companies realised that the allowances they had received at no charge were valuable assets, particularly in the midst of the financial credit crunch. The World Bank also reports that this sell-off was followed by a 'discernible increase in trading of EUA options (more calls than puts, on average), showing the intent of some installations to hedge any anticipated 2008–12 compliance exposure' (World Bank, 2009, p. 8). Such developments raise questions about the effectiveness of carbon trading systems in providing stable economic incentives to emitters who are assumed to respond rationally to price signals. If such schemes are, in practice, unable to deliver the stable and/or rising carbon prices that carbon-trading theorists deem necessary for long-term low-carbon investment decisions – the internalisation of external costs – it is difficult to see why they should be preferred over taxation or direct environmental legislation.

'Additionality'

The EU ETS allows for offsetting credits from joint implementation (JI) and CDM projects. Climate change mitigation is expected to be done at the lowest cost due to variance in the incremental costs of preventing the emission of one extra tonne of CO_2 into the atmosphere across countries. In the case of the CDM, a company from a developed country can invest in a project that is expected to reduce emissions in a developing or threshold country.[3] The company receives the tradable certified emission reductions (CERs) in return for this engagement; these are subsequently used to meet the CO_2 emissions benchmarks in their country of origin.[4] The *sine qua non* for CDM schemes to work is that they initiate projects to increase climate protection in developing countries in 'addition' to what would have happened without these projects, because, if certificates are issued for projects that would have been carried out anyway, developed countries increase their emissions without any simultaneous emissions reduction in the CDM guest countries. The CDM Executive Board, an agency specifically set up by the United Nations (UN) for the approval of such projects, issues allowances based on the assumed difference between the hypothetical CO_2 emissions in absence of the CDM project – the baseline – and the amount of emissions under consideration of the project (Witt and Moritz, 1998, pp. 95–9). The development of methods for defining 'additionality' and for evaluating the emissions saving effect from these projects has hitherto been one of the CDM Executive Board's most time-consuming and disputed activities (Trexler et al., 2006). Evaluation studies show that 40 per cent of currently registered CDM projects cannot prove that they would not have been carried out anyway. These projects represent 20 per cent of all CDM credits (Schneider, 2007). Manipulation and fabrication of data on the part of project applicants in order to achieve the required results were far from being exceptions to the rule;[5] and this does not appear to be just an initial problem, but a feature inherent to a system that encourages project applicants to generate as many as possible certificates for the lowest possible costs.[6] The European Commission's 2009 EU ETS Amending Directive will double the amount of CO_2 emissions that can be accounted for as JI and CDM, so that such credits can be used for up to 50 per cent of EU-wide CO_2 reductions by 2020. Given the consistent methodological difficulties around the notion of 'additionality', the widespread distorted use and, in many cases, outright abuse of the CDM, this decision cannot be supported. Its immediate consequence is that more CO_2 can be emitted from the EU territory than previously, while it remains unclear whether this surplus in emissions will indeed be compensated for by CDM projects outside Europe.

Bureaucracy and costs

Another alleged virtue of carbon-trading schemes put forward by neoclassical environmental economists is in the notion that its operation requires less bureaucracy and has lower costs than, for example, a tax on greenhouse gases. However, any empirical analysis of actual carbon-trading schemes demonstrates that the amount of administrative work necessary for the implementation and running of an entire new set of institutions at global, European and national levels in order to initiate and deal with the commodification of CO_2 is significant, weakening this proposal considerably. The tasks of these institutions include the measurement of emissions, the issuing of emissions rights, the registering of trades and trade platforms, the regulation of property rights, the validation and approving of CDM projects, enforcing compliance, ensuring and processing reporting and dealing with the widespread opportunities for fraud. Each EU member state sets up a special government agency to monitor the scheme. In Germany, this agency employed 120 people full-time in 2008, as well as a large number of freelance experts (Fisahn, 2008, p. 63). In validating CDM projects, a major concern is the impartiality of evaluators, who are accredited and listed by the CDM Executive Board as Designated Operational Entities (DOEs). These first identify the baseline or 'business as usual' scenario and then subtract the greenhouse gas emitted under the project scenario from the baseline. The result is the amount of emissions saved. However, in the EU ETS, it is the project applicant who assigns a particular DOE for a proposed project from the list provided by the CDM Executive Board. These rating agencies develop a range of CDM projects and are always looking for follow-up orders (Witt and Moritz, 2008, p. 95). There is an analogy here with the housing market, where rating agencies positively evaluated a range of non-viable mortgage products before the onset of the 2008 crisis (thereby contributing to its outbreak). Similarly, there is significant systemic pressure to produce positive evaluation outcomes within the EU ETS scheme, since the likelihood of being named as DOE by companies, again, depends significantly on their evaluation the last time around.[7] The costs for the substantial administration of carbon-trading schemes have hitherto been borne by the general public (the taxpayer) while the CO_2 emitting companies have contributed nothing so far. In Germany, energy clients paid twice: first as taxpayers for the set-up and maintenance of the new agency, and then as purchasers of energy, since energy companies passed on the price for allowances to customers despite the fact that they had received them for free – a 'socialisation of costs and a privatisation of profits' (Fisahn, 2008, p. 64, my translation). We can only speculate about the level of bureaucratic impact and the additional cost for taxpayers that an expansion of carbon-trading systems to the entire globe would entail, but there is no doubt that these would be considerable, given that most countries – including major emitters such as China or Russia – currently lack the necessary monitory equipment (Lohmann, 2006, p. 98).

Fossil path dependency

The final argument is that carbon markets help to introduce growth strategies based on renewable energies both in the developed and – via the CDM – in the developing world. This is also unlikely, given the experience of the EU (see Chapter 13) – and also from a theoretical standpoint. Contrary to neoclassical beliefs, carbon markets trigger a 'lock-in' effect on existing technologies, which is reinforced by the possibility of outsourcing the developed world's CO_2 reduction duties towards developing countries. Part II dealt with the processes by which technologies and production processes based on fossil energies gained a head start on other technologies and energy forms in the nineteenth century. US fossil-dependent technologies created a base of skills, research and resources that guaranteed their rapid growth and moulded societal production and consumption norms, including forms of mobility and transport and the creation of new lifestyles. During the process, alternative energy sources, technologies and means of transport were out-competed – on many occasions, thanks to the helping hand of a generous subsidy-providing state. What George W. Bush called an 'addiction' to oil and fossil fuels towards the end of his presidency has been labelled, in more academic terms, the 'entrenchment' of an energy regime 'in far-reaching technological, political and cultural webs' – a regime that gives these technologies and production modes 'the advantage of economies of scale, synergies with other industries, access to policymakers, accumulated specialist expertise, and subsidies of various kinds' (Lohmann, 2006, pp. 111–12). During the course of a long historical process, the bias towards fossil energy sources was inscribed or 'locked into' society's institutions, and indeed in its citizens' mental schemes, with the consequence that people think of fossil-fuelled technologies as being 'naturally' cheaper and more efficient than their alternatives. Hence, given the increased availability to cheaply outsource carbon emissions reduction duties via the flexible mechanisms, a transition towards an economy based on renewable energy carriers cannot plausibly be expected. And, even if CO_2 certificate prices were to rise, the technological lock-in effect and corresponding bias towards fossil fuels would be more likely to 'spur a search for more oil and gas than a search for better sources of energy' (Lohmann, 2006, p. 111).[8]

In relation to the developing countries, the evidence for the CDM as a means of spurring technological innovation and sustainable production is likewise weak. Witt and Moritz (2008, pp. 96–7) scrutinised the geographic distribution of CDM projects and found that, in 2008, 771 out of 1,033 registered projects were carried out in the four threshold countries: India, China, Brasil and Mexico. About one-third of CDM projects took place in India alone, while in China it was about one-fifth. The majority of the developing countries and, in particular the poorest ones, represent a much lower percentage of CDM projects than the four threshold countries and, therefore, have limited access or none to the intended technology transfer. In countries that have poor infrastructures and a weak rule of law, transaction costs are comparatively high, making CDM projects costly and risky enterprises from companies' point of view;[9] hence the vast majority of CDM projects are implemented in the so-called 'emerging markets', which also concentrate a great deal of FDI (Chapter 9). In these countries, the need to attract FDI provides policymakers with the 'twisted' incentive of not implementing far-reaching climate-protection legislation at the national level, since doing so would violate the very basis that attracts FDI in the form of CDM projects. These projects have to be 'additional', and that which already has legal status can hardly claim additionality. Thus, the lower the environmental standards, the greater the chances of authorisation for a proposed CDM project. In relation to the size of CDM projects, Witt and Moritz (2008, p. 98) argue that the transaction costs associated with the need to register a project tend to undermine small-scale projects embedded in regional economies, while supporting large-scale projects where such costs are comparatively low. Concerning the content of carbon-saving projects, the authors argue that most of these are classical 'end-of-pipe' solutions: rather than cutting the flow of raw materials into industrial processes, they tackle the problem after the resulting waste has already been emitted. Within the

CDM, 'marginal projects dominate, such as the containment of industrial gases by bolting on filters to already existing pipes' (Giddens, 2009, p. 190). Far from spurring technological change, carbon markets achieve CO_2 emissions reductions in a 'cumbersome and highly inefficient way' (Giddens, 2009, p. 190), if such reduction is achieved at all: Giddens (ibid.) calculates that half of the CO_2 reductions claimed as outcomes of the CDM are the 'result of "accounting tricks" and are empty of content'.

Finance-driven capitalism and climate governance

In conclusion, neither the empirical test of existing carbon-trading schemes nor the closer examination of its major theoretical proposals provides much cheer. Indeed, for understanding why free-market solutions to climate change are being sought, despite the fact that they do not address the issue adequately, one has only to look beyond the academic reasoning and consider wider societal transitions, including those of an ideological kind. Chapter 13 demonstrated that there are, in principle, three different ways of dealing with environmental issues in economic theory and political practice: environmental legislation, taxation and commodification. It is difficult to see why one of these methods should be more congenial to capitalism as a mode of production than others. Yet a regulation theoretical periodisation of capitalist development suggests that there is homology between capitalist accumulation regimes, modes of regulation and dominant environmental policy instruments. During Fordism, the state's role as proprietor or administrator of public goods and its capacity to act in areas such as economic, social and environmental planning was rarely disputed, and so were the corresponding 'top-down' government techniques in which Keynesian and Pigouvian tax regimes played a prominent part. During the crisis of Fordism and subsequent rise of finance-driven capitalism, the role of the state changed in two fundamental ways: on the one hand, its role as a proprietor and/or administrator of public goods weakened, partly as a result of privatisation policies in areas such as housing and energy, with the private sector playing an increasingly significant role. On the other hand, Fordist government techniques were complemented or substituted by governance models based on a distribution of labour between public and non-public actors (Chapter 12). This coincided with the neoliberal perspective becoming hegemonic in economic theory and policy making. The idea that market forces with the accompanying (re-)commodification and privatisation of public goods are per se superior to any kind of state regulation and public ownership became the dominant worldview and, as such, almost undisputable (Chapter 8). The new regime of dealing with environmental damages – and with climate change in particular – on the basis of the commodification of carbon emissions, of the artificial creation of private property rights and of emission trading as the main mechanism emerged as a part of this wider political and economic transformation. Indeed, it took no less than a neoliberal revolution to make policymakers see that their foremost *raison d'être* was to expand market regulation towards more and more areas in society. Where a market did not yet exist, it had to be created by a strong state.

In the light of these wider societal developments, it is far from coincidental that Voß (2007, p. 114) diagnoses a 'fundamental transformation of basic structures of environmental governance', within which tradable permits and certificates of various kinds became state of the art. And, like all major social transformation processes, the one from state regulation to market steering in environmental governance led to the emergence of new interest groups, conflicts and power relationships. The artificial creation of CO_2 as a commodity by governments fashioned new professional roles for groups such as lawmakers, accountants, lawyers, surveyors, consultants, journalists and engineers. Further indispensable and mutually reinforcing pillars of this emerging climate change governance edifice were the new public agencies that monitor the emission trading instruments at different regional, national and global levels, trading departments in companies, auditors for emissions, think tanks, consultancy firms and new kinds of banks and stock exchanges. New career paths opened up for project developers and website operators, which spread information on available investment opportunities and services according to the CDM. Ingenuous banks and investment firms developed new financial products of different risk categories. This range of different economic and political actors is united by the primary and very material common interest that this market emerging around the commodification of CO_2 emissions certificates continues to exist and to expand (Ptak, 2008, p. 48). And, compared to this interest, the question of whether carbon trading contributes anything to climate protection is indeed secondary. Due to the significant influence that these private interest groups and state bureaucracies mobilise in the public domain, it seems that the growth of the volume of emission trading as a lucrative field of investment and speculation is, of itself, meaningful instead.

The celebrated neoclassical presuppositions – *homo economicus*, total transparency (the idea that the same information on market processes is available to all market participants) and equal and unlimited market access for economic actors – might well be useful for certain purposes in economic theory. However, as an adequate understanding of the current state of the international division of labour and its mode of regulation, of which the new market for carbon emission trading is a part, they are misleading at best. Like all markets, the current world market is the product of far-reaching political and economic decisions that were taken by the governments of the most developed countries during the crisis of Fordism (see Part III). In the early twenty-first century, the world market is characterised by enormous and ongoing social asymmetries between developed and developing countries (see Part IV), as well as by increased competition between locations, which provides space for powerful financial actors to influence regulation on all levels. Hence, depending on an individual's country and place of origin, his or her potential to receive the necessary economic and cultural capital to obtain information on markets and to access and use them for one's personal benefit – and, more generally, the chances of prospering in or suffering from the existing global economic regime – are unequally distributed across countries and largely decided at birth. The omission of the issue of power and social inequality makes the neoliberal perspective, with its focus on procedural and technical aspects and its illusion that all players have the same information and power resources available, unsuitable for understanding the current governance process of climate change. Instead it helps to conceal the socio-historical causes of climate change as a global issue. The Kyoto discourse has sidelined 'the principle of inter- and intra-generational equity and responsibility' and has become skewed towards 'minimizing the burden of implementation on polluter industries and countries, instead of giving priority to the vulnerabilities of the communities [...] and countries at greatest risk and disadvantage'. The governance regime's primary focus is on the 'management of global carbon trade and meeting short-term targets, distracting due attention from the long-term challenge of stabilizing atmospheric greenhouse gas concentrations' (Najam et al., 2003, p. 223). Worse still, the system's existence creates the deceptive appearance that 'something is being done about the issue'. By assuring that tackling climate change as an issue does not contradict finance-driven capitalism and that this issue is dealt within its institutional structure, resistance and the establishment of alternative ways of working and living become more difficult. On top of allowing corporations and associated governments to manage climate change at the lowest financial cost and to open up a range of new career and investment opportunities, the existing climate change governance edifice has a detrimental impact at the individual level, where it undermines a transformation of the fossil consumption norm. Carbon-offsetting schemes are the pendant to the CDM for the individual consumer and offer a comfortable way of salving one's guilty conscience by maintaining the illusion that climate change can be dealt with or without the need for behavioural change.

Concluding Remarks

The scientific consensus on climate change is clear-cut. To avoid its worst effects, atmospheric CO_2 concentrations need to stabilise around or below 450 ppm (parts per million). Though greenhouse gas emissions need to peak before 2020 and then rapidly to decrease in order to meet this benchmark, no substantial steps are being taken to deal with this issue. Each year that goes by without transition to a qualitatively different energy pathway makes the 450 ppm goal less likely to be achieved. Hence, there is a real possibility that the present generation will put the planet on course for uncontrollable climate change and for a temperature rise of 5°C or more. In this case, future generations are likely to regard the present one as selfish at best. Yet it cannot be excluded that what is today called the 'Western way of life' will earn our generation a place in history books on a par with mass murderers: people who committed a crime against future generations of humanity. I wrote this book in order to illustrate the gap between the scientific knowledge of climate change and the lack of adequate action on the part of political, economic and environmental decision-makers. To achieve this aim, I have interpreted climate change as a social issue, with particular emphasis on its parallels to capitalism as a powerful social structure in contemporary society. My point of departure was a general analysis of the material and energy aspect of economic activity and of the structural tensions that characterise the relation between the conditions of reproduction of the capitalist economy and those of the reproduction of the ecological system. Subsequently, the book examined the most recent capitalist growth strategies in relation to their regulation of nature in general and of climate change in particular by applying key concepts from the regulation approach to Fordism and finance-driven capitalism. The last part of the book specifically dealt with the international regulation or global governance of climate change.

Part I has demonstrated that it is an anthropological constant that human beings need to interact with nature through labour for their subsistence. Labour processes, in turn, are always bound up with an irreversible and linear transformation of organic matter and energy. While the neoclassical perspective focuses on the circularity and reversibility of the 'return' of value and capital, thermodynamic economists stress the fact that any economic activity involves physical flow and throughput of matter and energy and that the earth's reservoir of natural resources is limited (Chapter 1). This is also at the heart of Marx's work, who – far from disregarding natural laws – made the pivot of his critique of political economy the dual nature of commodities, as constituting both exchange value and use value, and the dual nature of work, as creating both abstract value and concrete products through the transformation of raw materials and energy. Understanding this 'double character' provides insight, not only into further economic categories and the associated social relations at the level of the capitalist mode of production, but also into the corresponding tensions between the capitalist economy and the ecological system (Chapter 2) that amplify the greenhouse effect. More particularly, the further analysis of the dual nature of commodities and labour suggested that labour processes, under capitalist auspices, not only create abstract exchange value, but that they must also be understood as concrete stocks of invested time- and place-specific assets of matter and energy. Money and capital are qualitatively homogenous, quantitatively unlimited, divisible, mobile and reversible according to their exchange-value moment, while their energy and matter, or use-value aspect, is associated with qualitative heterogeneousness, quantitative limitations, indivisibility, locational uniqueness and irreversibility (Burkett, 2005). The 'conditions of production' – such as land, raw materials, fuels and other uncultivated resources – are used as 'free gifts' from nature and as sources of rents from the standpoint of valorisation, while their reproduction remains dependent on the – increasingly degraded and hence expensive – preservation of scarce natural resources (use-value moment). The production and accumulation of capital (exchange-value moment) is, due to the contradiction inherent in relative surplus production, linked to the expansion of the scale of production. Greater efficiency in the use of constant capital has hitherto been overcompensated by rising demand for natural resources. Hence, the expansion in scale has so far been translated into the increasing throughput of raw materials, auxiliaries and so on, especially fossil fuels, thus disrupting the carbon cycle and exacerbating climate change. The logic of the turnover process of capital forces entrepreneurs to reduce distances in time and space (exchange-value moment) while work processes and the associated linear and irreversible consumption and transformation of matter and energy always take place under specific temporal and local conditions. Capital's 'expansionism' tends to be accompanied by the degradation of the environmental conditions of production, and especially by reductions in their ability to act as sources and sinks for the permanently increasing flow and throughput of matter and energy. When these sources and sinks cease to function, their decelerating effect on the greenhouse effect is nullified, thus increasing the risk of negative feedback mechanisms within the climate system. Yet, under capitalism, the degradation of the environmental conditions of production is normally not addressed directly but makes itself felt indirectly in rising costs on the supply side (O'Connor, 1988) or by the *faux frais* of production (Marx), which any rational employer will attempt to pass on to the public.

The structural tensions between the principles of reproduction of exchange value and capital and of use values and nature can lead to environmental imbalances and ecological disasters, such as uncontrollable climate change. These tensions need to be managed and regulated in such ways that the emerging ecological imbalances may not deteriorate even further. However, like the social tensions arising from the unemployment and inequality that accompany capitalist development, environmental imbalances have taken different forms in different growth periods. The regulation theoretical perspective taken in this book (Chapter 3) stresses that the regulation of nature does not always proceed in the same way, but actually in different ways, which must themselves be understood as the temporary results of social struggles between various social actors. The choice of certain environmental policies or of a particular policy instrument over others is not in the first place a matter of having the 'better argument', but reflects wider societal power relations and asymmetries, including divisions within the capitalist class, as well as in the institutional traditions of different countries.[1]

From its origins in the Model T and the $5 day, Fordism was more than just a new accumulation regime in which mass production was cleverly combined with mass consumption. The associated exercise in social engineering that led to a qualitative change of workers' consumption practices (Chapters 4 and 6) was equally important. Originating in Ford's automobile factories and implemented by his 'Sociological Department', the new standardised consumption norm and code of behaviour for 'decent' employees was subsequently extended to the entire Atlantic space in the postwar period (Chapter 5). Ever more spheres of life beyond the working day became commodified. Social practices such as suburban housing and individualised automobile transport became the norm (with friendly assistance from major automobile corporations in the case of the US). The Fordist mode of societalisation was also based on a commercialisation of services and practices that were previously provided in non-capitalist ways. Since these had now to be bought as commodities, a regular – and possibly high – income became necessary for socio-cultural inclusion, while also being compensation of a kind for the physical injuries and psychological stress that individuals experienced in the Taylorised work processes. Far from being 'individual' and 'independent', my analysis has pointed to the social genesis and embeddedness in the Fordist consumption norms of 'choices' in the private sphere in areas such as housing or transport. Chapter 7 dealt specifically with the fossil-fuel nature of Fordism's energy regime: the geographic expansion of mass production and consumption went hand in hand with increasing throughput and a growing dependency on fossil-fuel energy sources such as crude oil. Indeed, one of the main reasons for the advantage of the US in economic development in comparison to Europe lay in the great availability and easy and inexpensive accessibility of fossil-fuel resources up until the first half of the twentieth century. As a corollary, the emerging international division of labour and its institutional structure after the Second World War cannot be fully understood without considering the fact that the US had become a net importer of fossil fuel by the 1950s. The US was thus interested in establishing an international order where the extraction of the globally available raw materials could be carried out in favourable economic conditions. The result was an international division of labour between industrialised countries and developing countries, which mostly exported raw materials and were attempting to initiate development strategies according to the principle of import substitution. During the growth period of Fordism, the world's total primary energy demand rose tremendously, the Organisation of Economic Cooperation and Development (OECD) countries consuming over 60 per cent of the total amount. CO_2 emissions per capita nearly doubled, the OECD countries emitting almost two-thirds of the total amount in 1973, despite the fact that these countries made up less than one-fifth of the world's population. Thus, the increase in atmospheric concentrations of CO_2 and, consequently, the development of climate change as an ecological crisis were closely tied to Fordist growth and to its corresponding international division of labour into 'extraction societies' (Altvater, 1992) and industrialised countries; these were also the main benefactors of fossil-fuel intensive consumption practices, such as individual car use.

From the mid-1970s on, Fordism entered a crisis that ranged from the exhaustion of the productivity potential of economies of scale, through diversified demand structures for industrial products and the spatial reorganisation of work processes, to the questioning of Fordist growth and the associated mode of societalisation by new social movements. To the extent that the basis for economic development became smaller – especially the parallel growth of companies' profitability and that of real wages – the hegemony within economic theory shifted from Keynesian to neoclassical thought, which soon also began to dominate policy discourses. A new accumulation regime gradually emerged, in which the focus was moved from 'management–labour' to 'management–shareholder' balance (Stockhammer, 2008; Chapter 8). Privileging shareholder value forces companies to change management techniques in response to economic cycles, including increased demand for flexible jobs and employment contracts. Further hallmarks of 'finance-driven' capitalism are the deregulation of financial markets, which facilitates capital flows and foreign direct investment (FDI), as well as a transition towards volatile exchange rates. This leads to increased uncertainty and short-term economic strategies on the part of companies. As a corollary, productive investment expenditure tends to be restrained; it also becomes more risky, being carried out under the imperative of sustaining higher profit rates than could be realised through financial investment. Socio-economic regulation generally focuses on supporting financial markets, and this is reflected in the transition, in capital allocation, from productive to financial forms of investment. Markets generally play a greater steering role than during Fordism and are increasingly applied to areas where state regulation previously predominated. In relation to the consumption norm, wages and salaries are more often complemented by other mechanisms, such as direct equity holdings, often through the intermediation of pension funds or via mortgage-based borrowing. Prospects of profits on the financial markets thus directly influence individual decisions to save or spend. The downside is a rise in household debt, the servicing of which is difficult during recessions. Commodified consumption moved into new areas and, given the context of decreasing real wages since the 1970s, it was increasingly based on private debt. New types of loans and mortgages were developed for this purpose, creating the illusion that consumption was no longer limited by real wages (Chapter 10). At the same time, as the neoliberal reforms were being initiated in the Atlantic space, major Asian countries such as China and India opened up their economies and turned into attractive locations for FDI and into investment areas for transnational financial capital (Chapter 9). China's gross domestic product (GDP), for example, grew by nearly 10 per cent annually between 1980 and 2010, and this coincided with a massive expansion in the country's scale of production and commodity output. There is also growing evidence for a 'Westernisation' of consumption patterns in major threshold countries such as China, India and Brazil. All these developments – the financialisation and transnationalisation of investment, the opening up of ever more regions of the world to capital accumulation – led to a doubling of the annual world's total primary energy use between 1973 and 2007 (Chapter 11). Finance-driven capitalism further intensified the fossil energy regime, which it had inherited from the Fordist period, with over 80 per cent of the 2007 energy supply stemming from fossil sources. The total amount of global CO_2 emissions nearly doubled between 1973 and 2007. Corresponding to relocations of assembly lines and the generally greater importance of new industrialising countries such as China in the world economy, the percentage of these countries in the production of CO_2 emissions also rose globally.

A common feature of Fordism and finance-driven capitalism is that both amplified rather than constrained capitalism's long-term trend towards economic and geographic expansion. This applies not only in relation to their accumulation regimes, but also (and especially) with respect to the consumption norm and energy regimes. Regardless of the concrete form of its regulation, capitalist economic and geographic expansion has been fuelled mainly by the burning of fossil-fuel resources for over a century, amplifying the greenhouse effect and producing climate change. Yet Fordism and finance-driven capitalism differ in point of their respective modes of regulation. The analysis in Part IV points to a transition from Pigouvian tax regimes towards Coaseian free-market solutions in environmental policies – a transition that corresponds to the general shift from Keynesian demand management towards neoclassical or 'Schumpeterian' supply management (Jessop, 1999). Taxation as a policy instrument dominated during the postwar period of environmental regulation, with a general predominance of 'top-down' state government. As the role of the state changed towards a more steering and enabling one (Koch, 2008) and socio-economic regulation in general began to be carried out by emerging networks of public and private actors at various levels (regional, national and European), environmental governance came to include a range of actors and featured the enhanced role of the market. Yet, while studies into multi-actor and multi-level governance structures at the national level point to the importance of trust among actors (Mayntz, 2009), the establishment of trust in international climate negotiations is severely reduced by the persistence of enormous global inequalities that translate into different power resources and bargaining positions for developed and developing countries in climate talks (Chapter 12). The outcomes and procedures of internationally adopted climate policies reflect these power asymmetries and the competitive interests of major transnational corporations and governments in developed countries. Due to their power resources advantage, delegations from Western countries are able to define climate goals that are not far-reaching enough to enforce a transition from fossil fuel to renewable energy sources. The agreed procedures to meet these goals – carbon trading in general and the 'flexible mechanisms' of the Kyoto Protocol and the EU ETS in particular – constitute new investment opportunities for financial capital, but have contributed next to nothing to a reduction in the atmospheric concentrations of greenhouse gases to date (Chapter 13). Instead, free-market 'solutions' to climate change are riddled with inherent anomalies (windfalls arising from the over-allocation of emission certificates, the contested issue of 'additionality' and the significant bureaucratic costs, to name only three). Consequently, Chapter 14 moved into focus the homology between the 'flexible mechanisms' and, more generally, market steering in environmental economics and a finance-driven accumulation regime. Neoliberal economists and policymakers argue that 'market liberation' provides a solution not only to the recent problems of Western welfare economies, but also to environmental issues. The constant reiteration that the climate crisis will disappear as soon as the emission of greenhouse gases is commodified in a global carbon-trading scheme not only contradicts the available empirical evidence on existing trading schemes, it also produces the illusion that 'something is being done' about the issue. Both the clean development mechanism (CDM) and carbon offsetting schemes function like 'opium for the people', who are led to believe that lifestyle changes are not necessary in order to tackle the issue. In fact a lot of money could be saved and some of people's potential for critical reflection could be restored, if these schemes did not exist at all.

In short, within the framework of the current accumulation regime and its consumption norm, and on the basis of the current mode of regulation and energy regime, the type of changes in production and consumption processes that scientists regard as necessary to mitigate climate change will be impossible to achieve. Neoclassical environmental economists celebrate the market as the main steering principle within a flexible accumulation regime, and they continue to have high hopes in the technological solutions that unlimited competition will bring about. These hopes, however, have not been fulfilled. While there is some evidence for the relative decoupling of GDP growth and CO_2 emissions per unit of economic output in OECD countries, major emerging economies demonstrate increases in relation to their primary energy intensity. Contrary to neoclassical environmental economic thought, the empirical analysis in Chapter 11 again confirmed the Jevons' paradox, first proposed in the nineteenth century: that increased efficiency in the use of fossil fuels results in an increase in demand for them. This helps to explain why relative decoupling has not been accompanied by 'absolute decoupling' (Jackson, 2009) – that is, by a growing economy combined with an absolute decrease in the throughput of energy and matter and CO_2 emissions. In fact, the only occasions where absolute greenhouse gas emissions fell in recent decades were periods of major socio-economic crises such as in Cuba and Eastern Europe after the fall of the Berlin wall or, more recently, in the EU during the financial crisis. Yet, in the case of the EU, this 'progress' is bound to be of short term, due to the anomalies inherent in the existing carbon-trading scheme. The scheme's utter inadequacy to bring about a fall in atmospheric greenhouse gas concentrations could be best exemplified by the fact that companies are allowed to 'bank' their emissions allowances during the periods in which they emit less than their prescribed CO_2 amounts – that is, during an economic crisis. Instead of initiating technological innovation and of shifting towards renewable energy sources, such companies will simply use their temporarily redundant certificates as soon as there is an economic upturn and the scale of production grows again.

In assessing what would need to be done to mitigate climate change, the current climate crisis (in combination with the financial crisis) is in some ways comparable to the situation after the Second World War, when a complete renewal of capital–labour relations came on the agenda. The prewar international regulatory system had proved incapable of combining the emerging industrial paradigm of mass production with a compatible consumption mode; this was partly due to the fact that trade unions were excluded from decision-making in socioeconomic regulation. It became obvious that capitalism as a mode of production had lost much of its appeal in the run-up to, during and in the aftermath of the Second World War, when the German Christian democrats demanded the socialisation of the energy sector and of key industries in their *Ahlener Programm* of 1946. A policy U-turn of this extent is indeed necessary today. Since uneven capitalist development and the corresponding enormous global inequalities lie at the heart of the current stalemate in climate negotiations, the advances in global climate policy that could achieve a working global deal presuppose a reduction in global social inequalities. If catastrophic climate change is to be avoided, capitalist development, global inequality and the climate crisis would need to be addressed as a whole, as well the ways in which they are interrelated. Trust could only be developed at the international level by integrating the issues of social inequality and the ecological needs of the planet in global governance (Roberts and Parks, 2006). Such a new development perspective could quickly focus on overcoming the fossil-fuel energy basis of the previous two growth strategies. This, in turn, would presuppose a transition in production processes towards renewable, especially solar, energy sources (Scheer, 2007), as well as qualitative changes in the Western way of life. And this necessitates redefining much narrower limitations of the market as a steering and allocation instrument than currently is the case. Respecting the principles of ecological reproduction means that humanity can no longer afford itself the luxury of not questioning capitalism or the efficiency of markets in relation to social and ecological issues. It was, after all, during the historically relatively short existence of the capitalist mode of production that climate change emerged, worsened and became a threat to human civilization.

The theoretical discussion in Part I proposes that it is not set in stone that capitalism is necessarily dependent on fossil-fuel consumption. Yet, due to the lock-in effect of fossil fuels in previous capitalist development, other energy carriers have not had the required political and economic backing to accomplish the transition of production and consumption patterns to renewable energy sources, which is necessary to mitigate climate change. A sustainable growth strategy would, at the very least, involve the reconfiguration of social, state and market steering and the establishment of international institutions (not unlike the Bretton Woods institutions after the Second World War) powerful enough to limit and steer capital valorisation in accordance with ecological laws – and in particular to specify boundaries to greenhouse gas emissions for companies, countries and individuals, according to climate science expertise. Such institutions would need to be strong enough to enforce ecological standards in production and consumption in the developed countries and, if necessary, to abolish market steering, the profit motive and private property in areas where climate goals are not being met. A mixed economy with different forms of property (private, communal, societal, state property etc.) would result. In contrast to the present dysfunctional climate governance system, which largely reproduces the 'mistrust built on decades of unequal experience' (Roberts and Parks, 2006, p. 214), the longer-term interests of the developing countries would need to be taken seriously in any real 'global deal'. These interests have not qualitatively changed over the last decades, as Najam et al. (2003, p. 226) point out; they focus on the 'creation of a predictable, implementable and equitable architecture for combating global climate change that can stabilize atmospheric concentrations of greenhouse gasses within a reasonable period of time', and on 'enhancing the capacities of communities and countries to combat and respond to global climate change, with particular attention on adaptive capacity that enhances the resilience of the poorest and most vulnerable communities'. To restore trust and build a global low-carbon development strategy, the developed countries would need clearly to 'signal their commitment to this new "shared thinking" though a series of confidence-building measures' (Roberts and Parks, 2006, p. 217). Anything less than legally binding commitments in relation to domestic reductions in greenhouse gas emissions and substantial increases of support for mitigation and adaptation in developing countries would be seen 'as a failure to seriously invest in repairing the trust deficit' (Baer et al., 2008, p. 24). A further indispensable element for building trust at the international level would be the 'greater stake' of developing countries in the governance and decision-making of international financial institutions (Roberts and Parks, 2006, p. 24).

Enough scenarios exist today that illustrate the kind of changes that need to be implemented to transform dominant production and consumption patterns in ways that limit greenhouse gas emissions as climate science requires. For the purpose of the present book, only a few need be mentioned. In order to tackle global inequality and climate change in tandem, what is required is the provision of 'greater "environmental space" to late developers, supplying meaningful sums of environmental assistance, funding aid for adaptation and dealing with local environmental issues' (Roberts and Parks, 2006, p. 217). As these two authors convincingly argue, this would need to be complemented by abandoning the (de)regulation principle of facilitating FDI and, in particular, finance investment across the globe. Likewise, international 'agreements' like Trade-Related aspects of Intellectual Property Rights (TRIPS), Agreement on Trade Related Investment Measures (TRIMS) and General Agreement on Trade in Services (GATS) and Western agricultural and fossil-fuel subsidies, which all undermine the long-term interests of developing countries, would need to be phased out. These subsidies would need to be transferred to low-carbon energy sources and to the support of public transport. Further elements of a new environmental governance system would include strict regulation for setting efficiency and carbon use standards for buildings, vehicles, urban development and land use. Procedurally, the present analysis proposes that carbon taxes, 'green' taxes on material intensity (regarding the use of metals, water, wood, plastics and so on) and legal action would be more effective at steering the transition towards a low-carbon growth strategy than the market 'solutions' advocated by mainstream economists, presently applied by policymakers. Revenues from such taxes could be used to fund 'low-carbon energy and increase efficiency, or offer rebates to buyers of greener, more efficient equipment' (Lohmann, 2006, pp. 330–332).

Political re-regulation and state intervention are necessary not only on the demand side of energy use (via taxation and the setting of quotas) but also on the supply side. Massarat (2008), among many others, proposes determining and monitoring a gradual reduction, on the worldwide scale, of the global annual output of fossil energy carriers such as crude oil. If this reduction amounted to 2 per cent per year, there would be no more CO_2 emissions resulting from oil combustion 50 years after the beginning of the scheme. The enormous amounts of public money currently being squandered on carbon-trading schemes could be redirected towards supporting oil-exporting countries in transitioning their economies. In contrast to trading schemes, a legally binding 'keep the oil in the soil' approach would constitute a real incentive for companies to shift their energy supply towards renewable, especially solar, energy sources in the foreseeable future. Integrating multiple technological, regulatory and consumption-related approaches, the 'Greenhouse Development Rights' framework brings together socio-economic development and greenhouse gas emission reduction benchmarks and translates them into national action plans, including the structure of a climate tax for individual countries and a ranking of countries in relation to these benchmarks.[2] On the individual level, the G77 has proposed allocating pollution allowances on a per capita basis. Following a rather simple egalitarian logic, every global citizen would be allocated an equal entitlement to pollute the atmosphere according to specified individual emissions budgets, in accordance with the global greenhouse gas reduction benchmarks identified by climate scientists (Grubb et al., 1999, p. 270). In contrast to free-market or commodification models, this would unavoidably call Western consumption patterns into question – including carbon-emission-intensive practices such as individual flights and automobile use – since citizens of the developed countries have already used a disproportionate amount of their share of carbon and would, therefore, need to contract their carbon budget substantially. Citizens of developing nations, who have thus far emitted fewer greenhouse gases than their proportional share, could, in contrast, increase fossil-fuel consumption for a certain period and eventually converge with the developed countries (Roberts and Parks, 2006, p. 145).

Although such proposals on the action required to tackle climate change are essential – they demonstrate that a different economy and a qualitatively different set of consumption practices are possible – the present book has focused on the powerful and partially hidden social structures arising from capitalism and its two most recent growth strategies that contradict and undermine the implementation of these suggestions into practical policies. Our analysis suggests that what is often euphemistically presented as 'globalisation' – the formation of the global market – is a political creation, which, like the genesis of national markets, was not the simple result of the gradual extension of exchange relations but the outcome of mercantilist state policies that aimed at fostering the commodification of land, money and labour (Polanyi, 1944). And, just as the historical constitution of national markets cannot be understood without referring to the 'visible hand' of nation-states, the new universal model is not an automatic result of the laws of technology or of *the* economy, but must be understood by considering the power relations in operation at the international level. With hindsight, the term 'globalisation' does indeed seem to be a euphemism for the universalisation of a socio-economic system that served the interests of US elites especially well. Authors such as Bourdieu and Harvey point out the 'constraints' that finance-driven capitalism and the large transnational corporations impose on national policymakers and how the neoliberal perspective became generalised as a universal model of economy and society. This was a new *doxa* that Bourdieu understands to be the result of the symbolic labour of a new type of intellectual, often working in financially stable think tanks. It was not only in Chile that the new distribution of labour between the field of intellectuals and the field of power – between those ' "thinkers" longing for power and people in power longing for ideas' (Bourdieu et al., 2002, p. 182) – emerged, making possible the transition of the accumulation regime towards a finance-driven one. Like the Catholic *doxa* of the Middle Ages, the new neoliberal *pensée unique* seemed to provide solutions for all kinds of social and ecological issues. Often neutralised in academic terms and amplified by associated intellectuals within and outside the mainstream media, these solutions are relentlessly preached – not least to students who will constitute the elites of the future.

Notwithstanding the extraordinary economic, political and symbolical power of financial capital and allied intellectuals, an adequate understanding of the spread of neoliberal categories cannot be reduced to a notion of indoctrination and manipulation. The impact of neoliberal think tanks on policymakers only became so efficient because there existed a certain readiness for collaboration or – as in the case of women and their male dominators – a degree of practical consent on the part of those who are exposed to power and symbolical violence (Bourdieu, 1984b, p. 386). In fact, this submission of the dominated goes much further than the ideological distortions of the 'consciousness' emphasised in the Marxist tradition. The social structure is inscribed not only in the *ideas* of the dominated, in their mental representations, but also in their bodies, in their 'schemes of perception and dispositions (to respect, admire, love, etc.), in other words, beliefs which make one sensitive to certain manifestations, such as public representations of power' (Bourdieu, 2000, p. 171). It follows that we would succumb to a 'scholastic fallacy' if we expected heterodox practice and social change to be accomplished by 'raising of consciousness' alone. 'While making things explicit can help, only a thoroughgoing process of countertraining, involving repeated exercises, can, like an athlete's training, durably transform habitus' (p. 172). Yet, while one should not underestimate the potential role of intellectuals in bringing about heterodox discourse and practice, Bourdieu refers to a second precondition for this to occur, which is perhaps even more crucial. 'Prophetic' or 'heterodox discourse', such as that of a low-carbon and solidarity-based economy and society, has

> more chance of appearing in overt or masked periods of crisis affecting either entire societies or certain classes, that is, in periods, where the economic or morphological transformations of such or such a part of society determine the collapse, weakening, or obsolescence of traditions or of symbolic systems that provided the principles of their worldview and way of life.
>
> (Bourdieu, 1991, p. 34)

In other words, the chance of alternative ways of thinking and acting becoming hegemonic depends on the existence of a crisis and transformation of the 'objective' economic, political and cultural structures of society and of the corresponding symbolic systems and forms of habitus. Any intellectual critique of the neoliberal perspective and its application to environmental governance is only effective insofar as this 'countertraining' builds upon such crisis.

The present book points to a twofold crisis of finance-driven capitalism – in relation to its economic and political structure and in relation to the climate. The fact that the economic crisis of 2008 occurred is not entirely surprising, given the relative detachment of finance assets from real value creation after the liberalisation and deregulation of financial markets. While real wages stagnated, people were motivated to 'borrow and consume as if their incomes *were* improving' (Stiglitz, 2010, p. 2), to keep demand and consumption levels stable. Rising levels of debt were seen as unproblematic, since the gains from financial investments would spur both investment and consumption. However, an equity-based regime depends on monetary policies that control financial bubbles, since there is always the risk that the diffusion of finance may push the economy towards structural instability (Boyer, 2000). In the absence of such policies, the finance sector becomes self-referential; it was not only finance managers that happily cashed in enormous bonuses for their handling of financial transactions, for which they expected returns between 15 and 25 per cent after tax in relation to capital advance. However, such rates of capital valorisation were increasingly difficult to sustain, since financial investors were competing for a limited number of investment possibilities in the real economy, thereby exploiting ever narrower profit opportunities. In the short term, financial investors could maintain profitability through 'accumulation by dispossession' (Harvey, 2005) and in particular through the sale of public goods such as social housing, the privatisation of the energy and care sectors and the creation of entire new markets such as that for carbon emissions. Yet, in the long run, the balance between the finance sector and the real economy was lost, thus making the burst of the bubble unavoidable. The political and economic crisis of finance-driven capitalism is aggravated by the growing importance of countries such as China, India and Brazil in the world economy, the following of non-orthodox political and socio-economic growth strategies in several South American countries and, more recently, the revolutions in Northern Africa. Despite their many differences, these developments at the international level are moving in the same direction – these countries are no longer prepared to play the role of 'extraction economies' that provide the Atlantic space with cheap fossil energy, especially crude oil.[3]

The climate crisis, for its part, is constantly ignored or understated by policymakers – and used instead as an additional investment area for financial capital – because of the delayed reaction of the climate system to past and present excessive greenhouse emissions (see Introduction). Activists such as Larry Lohmann (2006, p. 337) are, thus, unequivocal in saying that 'dealing with the climate crisis is going to involve democratic political organising and an uphill struggle'. One of the experiences of the Copenhagen Climate Summit in 2009 was indeed that it seems futile to wait for decisions on 'global climate deals' at UN summits. The informal exchanges and the continued communication between activists that began at the Copenhagen Climate Forum were actually more important, suggesting that it is possible to combine political commitment with a significant reduction in one's individual carbon footprint, and that this need not necessarily mean a lifestyle shaped by austerity. In fact, partially as an individual response to the economic crisis, more people seem to be engaging in seemingly profane and parochial practices such as urban gardening, establishing of local food cooperatives and non-profit service networks, holidaying closer to home, using public transport such as railways and ferries and avoiding air travel. As a perhaps unexpected side effect for some, resisting the immense societal pressure to partake in endless, carbon-intensive consumption can indeed achieve an empowering effect, almost constituting an oppositional act.

Yet whether this twofold crisis of finance-driven capitalism will eventually result in the kinds of changes to the growth strategy outlined here and will actually lead to a carbon-free and solidarity-based economy and society is far from certain. This is because the crumbling of *doxa* – the process during which the structures of society that are normally taken for granted are being made explicit and objective, for everyone to see through – has, historically, only rarely led to its replacement by heterodox thought and practice. More often than not, the crisis of an established social order has resulted in a new kind of *orthodoxy* – from European fascism via Pinochet's *coup d'état* to more recent cases of religous extremism – where the social order is upheld and the dominant interests are defended by abolishing or weakening democracy and replacing it by authoritarian rule and the use of force. As the financial crisis evolves, new types of right-wing populist movements are celebrating electoral success all over Europe. These combine a conservative critique of financial capitalism with chauvinistic and xenophobic slogans and could, therefore, well provide the popular basis for an authoritarian 'solution' to the crisis – one in which the 'West' defends its way of life and the corresponding international division of labour using its military power, while closing and monitoring its borders and abandoning the victims of economic crisis and climate change in the developing countries to their fate.

Notes

Introduction

1. The location on earth where the greenhouse gas is released is not important for the production of the greenhouse effect: one tonne of greenhouse gases from Luxembourg causes exactly the same amount of damage to the world's climate as one tonne from China. Thus, climate change has been a truly global issue from the beginning.
2. The fifth report is expected in 2015. The main findings and data of the fourth report are summarised in the IPCC's synthesis report, which is available online at http://www.ipcc.ch/publications_and_data/ar4/syr/en/spms1. html (accessed on 20 January 2010).
3. The assertion that climate change is both real and caused by human activity is worth mentioning, since, until recently, the world was still debating whether or not climate change was taking place, and whether or not it was linked to human activity. While climate change scepticism continues to be a flourishing and lucrative industry, the scientific 'debate is over and climate scepticism is an increasingly fringe activity' (UNDP, 2007, p. 5).
4. While the IPCC scenarios often use the year 2100 as a reference point, it is important to remember that temperature adjustments to rising concentrations of CO_2 and other greenhouse gases are likely to continue to take place in the twenty-second century (UNDP, 2007, p. 35).

1 Nature and the Work Process

1. I will deal with neoclassical accounts of how to deal with climate change in detail in Chapter 13.
2. Daly (1985, p. 280) pointedly remarks that 'studying economics in terms of the circular flow without considering the throughput is like studying physiology in term of the circulatory system without ever mentioning the digestive tract'.
3. An *isolated* system is defined as being 'closed to *both* energy and matter transfers in or out, while a *closed* system is only closed to matter transfers' (Schwartzman, 2008, p. 46).
4. As Herman Daly (1985, p. 284) put it: 'To deny the relevance of the entropy law to economics is to deny the relevance of the difference between a lump of coal and a pile of ashes.'

2 Capitalism, Nature and Climate Change: A Structural Analysis

1. The concrete is concrete because it is the concentration of many determinations. In order to distinguish essential categories from more mediated

ones, Marx (1973, p. 148) proposed a distinction between research and pre-sentation: 'Along the first path the full conception was evaporated to yield an abstract determination; along the second, the abstract determinations lead towards a reproduction of the concrete by way of thought.' Marx's critique of German philosophy and its foremost representative, Hegel, is that he fell prey to the illusion 'of conceiving the real as the product of thought concentrating itself, whereas the method of rising from the abstract to the concrete is only the way in which thought appropriates the con-crete, reproduces it as the concrete in the mind. But this is by no means the process by which the concrete comes into being' (see also Koch, 2000, p. 24).

2. Similarly, in relation to the historical forms of the work process, regarded as being the essential exchange between humanity and nature, Marx argued that it 'is not the *unity* of living and active humanity with the natural, inorganic conditions of their metabolic exchange with nature, and hence their appro-priation of nature, which requires explanation [...], but rather the *separation* between these inorganic conditions of human existence, a separation which is completely posited only in the relation of wage labour and capital' (Marx, 1973, p. 489).

3. In contrast to the 'first' contradiction, which refers to capitalism's tendency to produce increasing social inequality in terms of income and economic wealth; this limits profitability and results in a crisis on the demand side, since commodities in ever-increasing numbers cannot be sold. I would, how-ever, not use the qualification 'secondary' contradiction, since it suggests a hierarchy of tensions or 'contradictions' in the relationship between cap-italism and nature. By contrast, this chapter shows that this relationship is caught up in a whole range of structural tensions or 'contradictions' (Table 2.1). It is difficult to see why one of these should be given primacy over others.

3 The Regulation of Nature and Society in Different Capitalist Growth Strategies

1. Dietz and Wissen (2009) identify a weakness in the structural 'eco-Marxist' perspective in that the focus on the natural limits of capitalism does not sufficiently consider the conflicts and struggles that arise from the capitalist appropriation of nature in its concrete manifestations 'on the ground'. These do not normally take the form of 'great crises', but they are nevertheless rele-vant for an emancipatory perspective: the day-to-day struggles around issues such as the privatisation of water supply or the already disastrous effects of climate change in developing countries.

2. See Koch (forthcoming) for a discussion of the expressions 'mode of produc-tion' and 'social formation' in Poulantzas.

3. DESERTEC aims at promoting the generation of electricity by using solar energy and wind energy in the deserts worldwide. This concept will be imple-mented in North Africa and the Middle East by the consortium DII GmbH, formed by a group of European companies and the non-profit DESERTEC Foundation.

4. Bourdieu's sociological analysis of the links between taste and social structure includes a rupture with Kant's 'pure aesthetics' and his distinction between 'what pleases' (*was gefällt*) and 'what amuses' (*was vergnügt*) and points to the social genesis of the different variants of taste and, in particular, to the correlation between taste and social class. By measuring different forms of cultural practices – like eating habits, styles of dress and the appreciation of art – he brings all the acts of delimitation and valuation into systematic order. If 'vulgar' taste emphasises the practical function of a piece of art, 'better' taste stresses its aesthetic form; if ordinary taste is led by quantity, 'pure' taste favours quality; if in popular taste matter and substance dominate, the cultural avant-garde emphasises the style (Koch, 1996 and 1998a).
5. These symbolic activities can turn out to be quite stressful and costly, as recent British research on social practices such as children's birthday parties (Clarke, 2007), weddings (Boden, 2007) or the furnishing and décor of children's bedrooms (Cieraad, 2007) have shown.
6. Elaborating on this, other authors have linked the regulation approach to the sociology of Pierre Bourdieu, and in particular to the concept of habitus (Koch, 2003, pp. 38–43; Herkommer, 2004).

5   The Geographic Extension of Fordism

1. Despite the fact that this expansion was, at this initial stage, to the detriment of investment and employment in the US (Musgrave, 1975).
2. Maddison (1995, p. 105) remarks that the boom was largest in countries that had suffered most from the 'protectionist, dirigiste, and otherwise defensive policies in the interwar years'.

6   Mode of Societalisation and Consumption Norm

1. For such a comparative quantitative study, see Koch 2006a (Chapter 5).
2. It is no coincidence that Bourdieu uses a dictum formulated by Bertrand Russel as the epigram for his book *The Social Structures of the Economy*: 'while economics is about how people make choices, sociology is about how they don't have any choice to make'.

7   A Fossil Energy Regime

1. Twelve countries currently belong to OPEC: Algeria, Angola, Ecuador, Iran, Iraq, Kuwait, Libya, Nigeria, Quatar, Saudi Arabia, the United Arab Emirates and Venezuela.
2. The *tonne of oil equivalent* (toe) is the amount of energy released by burning one tonne of crude oil. Multiples of the toe are often used, in particular the megatoe (Mtoe, one million toe) and the gigatoe (Gtoe, one billion toe).
3. CO_2 emission values presented here do not include emissions from land use change or emissions from bunker fuels used in international transportation.

8 The Rise of a Finance-Driven Accumulation Regime

1. See Valdés (1989) and Osorio and Cabezas (1995) for the influence of the Chicago School of Economics on Chilean economists.
2. Neoliberalism's breakthrough as a school of academic thought came with the award of the Nobel Prize in economics to von Hayek in 1974 and to Friedman in 1976.
3. Robert Lucas and Ben Bernanke, both senior figures in the US FED, were convinced of the stability of this new global financial edifice. Towards the end of the 1990s, they went as far as to claim that, 'while the economy would continue to suffer from occasional setbacks, the days of really severe recessions, let alone worldwide depressions' were 'behind us' (cited in Krugman, 2009, p. 22): a bold statement, given that global capitalism had featured 124 minor and major crises between 1970 and 2007 (Laeven and Valencis, 2008), never mind the greatest financial and economic crisis since the Great Depression that began in 2008.

9 The Recomposition of the International Division of Labour

1. In Germany, a great part of the media continues to celebrate the country's status as the 'export world champion'. However, Germany's economic success is largely dependent on its ability to export much more than it imports, while domestic demand has weakened as a result of stagnating real wages over recent decades. This overdependence on the export sector became obvious during the financial and economic crisis and the corresponding massive export deficits. Since domestic demand could not counter-balance them, the country experienced negative growth in 2009.
2. This integration found a preliminary culmination in 2001, when China joined the WTO.
3. It is estimated that about 10 million Indians lived here 500 years ago, before South America became part of the capitalist world system.
4. All data are taken from 'Rainforest Facts': http://www.rain-tree.com/facts.htm (accessed on 20 January 2010).

10 A Worldwide Consumption Norm (Based on Debt) and the Financial Crisis

1. I experienced this development at first hand when I moved to Northern Ireland in 2000. In Germany, young people were only allocated a credit card if they could also provide proof of a steady income, usually based upon permanent employment. In contrast, in Northern Ireland, I was astonished to discover that most of my students not only had one but very often several credit cards. Many were experts in moving credit card debts from one credit card company to another, thereby saving fees. Given the substantial tuition fees at institutions such as Queen's University Belfast and the comparably

high costs of living in Northern Ireland, these students quickly built up debts in the magnitude of GBP 30,000 or more.

2. According to Herman Minsky, in Ponzi finance, expected income flows do not even cover interest cost. As a result, a company must borrow more or sell off assets simply to service its debt, in the hope that either the market value of assets or the income rises sufficiently to pay off interest and debt.

3. Political decisions such as the repeal of the Glass-Steagall Act are viewed as major contributing factors to the crisis. 'As the shadow banking system expanded to rival or even surpass conventional banking in importance, politicians and government officials should have realized that we were re-creating the kind of financial vulnerability that made the Great Depression possible' (Krugman, 2009, p. 163).

4. Andrew Simms (2005, p. 95) calculates that, if 'among all the world's people we were to share equally a safe volume of greenhouse gas emissions, a single long-haul flight would take up one person's entire ration for several years'.

5. Since the origins of Fordism, there has been a highly unequal distribution of vehicle ownership. In the mid-1990s, there were 750 cars for every 1,000 US citizens, while in China there were just 8 for the equivalent number of people, and 7 in India (Simms, 2005, p. 129). The recent catch-up developments in these countries have changed this picture somewhat.

11   The Globalisation of the Fossil Energy Regime

1. On the basis of IEA data, I calculated these values by dividing the amount of CO_2 emissions by the number of the world population for 1973 and for 2007, respectively (Tables 7.4 and 11.4).

2. This echoes reports on wartime Britain, according to which life expectancy during this period grew much faster than in any period before or after (Sen, 1999).

12   Multinational Governance in an Unequal World: The Kyoto Process and the Actors Involved

1. Central features of 'de-democratisation processes by governance' (Dietz, 2009, p. 200) include the selectivity of actors and the corresponding exclusivity of arenas of governance. As Dietz argues, this is especially the case where only those actors have access to negotiations and decision-making who dispose of expert knowledge, sufficient financial means, the backing of influential organisations and the command of English as the international conference language.

2. Inequality is much higher at a global level than in Western Europe. In Germany, for example, the Gini coefficient has moved between 25.7 and 27.0 over the last three decades, while in Sweden, one of the most egalitarian countries, it ranged between 19.7 and 25.2 (Koch, 2006a, p. 65, 85).

3. The same, if to a lesser extent, applies to India (Milanovic, 2005, p. 91).

4. Milanovic (2005, p. 204) found that income inequalities in Indian states and Chinese provinces increased by one half and one quarter, respectively, between 1980 and 2000.

5. Milanovic (2005, pp. 103–35) calculates the global Gini from national household surveys for the years 1988, 1993 and 1998. His sample comprises about 96 per cent of the world's population.
6. Dietz (2009, p. 203) studied the case of Nicaragua, where the entire ministry of environment consisted of half a dozen employees in the early 2000s. At that time its financing was totally dependent on external sources.
7. Only the developed countries are normally able to provide a reliable state under the rule of law that guarantees private property, the principle of equivalence in exchange relations and the legal security of economic subjects.
8. For the period until 2008, I largely follow the account of events and developments in Brunnengräber et al. (2008, pp. 87–96).
9. Again, I make special reference to Brunnengräber et al. (2008, pp. 97–109).

13 Theory and Practice of Carbon Emission Trading: The Case of the EU ETS

1. Coase was a professor at the University of Chicago, and, like Friedman, was awarded the Nobel Prize in Economics in 1991.
2. These can be combined in a concrete empirical case.
3. There are also positive external effects, where the public benefits from an activity that the market undersupplies. Pigou (1932) suggested subsidising such positive externalities – education, for example – by the state.
4. In defense of Pigou, one could, however, question whether the lack of absolute knowledge of the gaps between private and public costs justifies the fundamental rejection of eco taxes by free-market economic environmentalists; even in the absence of this knowledge, the existence of the tax is likely to influence economic activity – and thereby to increase market efficiency.
5. According to Coase (1960), 'transaction costs' refer to the costs of the operation of the pricing system.
6. The 'operation of a pricing system is without costs, private exchange of property rights will lead to efficient resource allocation' (Coase, 1960, p. 2).
7. It was partly due to Stern's authority that many economic and political circles changed their mind on climate change and began to agree with his main argument that doing nothing about the issue will turn out to be more expensive and to impede economic activity more than taking action (Stern, 2007). Thus, unwillingly, Stern provides further proof for O'Connor's observation that, under capitalist auspices, the degradation of the environmental conditions of production makes itself felt only indirectly – as rising costs that diminish profits (O'Connor, 1988; Chapter 2).
8. Altvater and Brunnengräber (2008, p. 10) point out the irony that characterises the commodification of CO_2 – a 'non-value' that people normally try to get rid of.
9. These include not only private companies, but also the House of Representatives. If members reduce their emissions to a greater extent than agreed, they can sell dispensable allowances or 'save' them for the future.
10. It is nevertheless paradoxical that the US played the crucial part in ensuring that the CDM became an important element of the Kyoto Protocol – as this was the country that later refused to ratify the Protocol.

11. These credits are labeled Emission Reduction Units (ERUs) in the case of JI projects, and Certified Emission Reductions (CERs) in the case of CDM projects.
12. The European Employment Strategy, for example, is operated on the basis of a similar procedure (Ashiagbor, 2005).
13. According to Gilbertson and Reyes (2009), more than 80 million tonnes of carbon offsets were bought as part of EU ETS in 2008.
14. The EU Commission (2008b) intends this percentage to rise to 60 per cent in the course of the third trading period.

14 The Flaws of Free-Market Solutions for Climate Change Prevention and Their Homology to Finance-Driven Capitalism

1. This attitude is likely to be amplified by the fact that a global post-Kyoto agreement on CO_2 emissions reduction is currently unlikely to be reached. The same seems to be true at regional level: the US senate refused to introduce a carbon-trading scheme, which could have linked up with the EU ETS, in July 2010.
2. A further factor in this consideration is the likelihood that such fees are actually imposed by state authorities. This is likely to vary across countries.
3. Like the price for carbon emission certificates, the CDM market has turned out to be profoundly volatile. As a result of the economic crisis, confirmed transactions for primary CERs declined by nearly 30 per cent in value between 2007 and 2008 (from 552m CERs to 389 million). In the same period, the confirmed transactions for JI also declined by 41 per cent in value (World Bank, 2009, p. 8).
4. Primary CERs are purchased directly from entities in developing countries and are linked to participation in a CDM project, while secondary CERs are purchased independently from the participation in a project that reduces the emission of CO_2 in a developing country.
5. In relation to India, Michaelowa and Purohit (2007) point out that about every third UN-registered CDM project could not furnish plausible proof of additionality. Haya (2007) comes to a similar conclusion with regard to the building of hydroelectric power plants in China.
6. Especially lucrative in terms of CDM credits are projects proposing the destruction of trifluoromethane (HFC-23) (Witt and Moritz, 2008, pp. 99–101). The emission of one tonne of this ozone-depleting refrigerant gas has the equivalent harmful effect on the atmosphere of 11,700 tonnes of CO_2. HFC-23, which is emitted during the production of the refrigerant chlorodifluoromethane (HCFC-22), can be easily removed with the help of cheap gas scrubbers, and it costs less than one euro to destroy one tonne of CO_2 equivalents this way. According to Wara (2007), the projected total cost of destroying HFC-23 via carbon credits will amount to 4.7 billion euro by 2012, even though installing the necessary gas scrubbers costs less than 100 million euro. This enormous profit margin creates a perverse incentive for companies to expand refrigerant plant capacity via the CDM – a violation of the Montreal Protocol, which states that industrialised countries have to give

up this type of production by 2020 in order to protect the ozone layer. The case demonstrates not only that economic incentives arising from the CDM are often contradictory to climate change mitigation, but also that different instruments in environmental policy can actually cancel each other out.

7. Schneider's (2007) analysis points in the same direction, highlighting high competitive pressure among DOEs and relatively low and falling prices for evaluations and tight schedules.

8. Fossil path dependency and the interest in dealing with CO_2 emissions within the dominant economic logic – through commodification and carbon trading – is also expressed in the great hope that leading policymakers put in carbon capture and storage (CCS) – a technology that will remain in the testing stage for many years to come. While the companies developing it receive huge amounts of government funding, it is at present entirely unclear as to whether it will be operational in the future at all.

9. The geographic concentration of CDM in a handful of threshold countries indicates that the CDM should not be seen as an instrument of 'development aid'. It would be dangerous – but indeed not unlikely, in the light of the ongoing free market devoutness amongst policymakers – to conflate the two by abolishing or cutting 'traditional' direct aid for the sake of the 'flexible mechanisms'.

Concluding Remarks

1. Comparative analysis of the diverse institutional preconditions for establishing a low-carbon growth strategy will be focused upon in future research. Chapter 11 has demonstrated that Scandinavian countries – though still emitting more greenhouse gases into the atmosphere than recommended by the International Panel on Climate Change (IPCC) – emit considerably less than other capitalist countries and use renewable energy sources more often. These preliminary comparative results point to synergies between the levels of CO_2 emissions and varieties of market coordination, traditions of welfare capitalism and post-Fordist development paths in OECD countries that should be further explored.

2. See, for further details, http://gdrights.org/wp-content/uploads/2009/12/Principle-based-A1-targets-draft2-3.pdf.

3. In fact, the social changes in Northern Africa of 2011 alone could result in an even greater increase in oil prices than those of the 'oil shocks' of the 1970s, constituting an 'objective crisis' (in Bourdieu's sense) in finance-driven capitalism's fossil fuel energy regime.

References

Aglietta, M. (1987) *A Theory of Capitalist Regulation: The US Experience*, 2nd edition, London: Verso.

Aglietta, M. (2002) 'The International Monetary System', 64–72, in Boyer, R. and Saillard, Y. (eds), *Regulation Theory. The State of the Art*, London: Taylor and Francis.

Agnew, J. (1987) *The United States in the World Economy*, London: Cambridge University Press.

Altvater, E. (1992) *Der Preis des Wohlstands oder Umweltplünderung und neue Welt(un)ordnung*, Münster: Westfälisches Dampfboot.

Altvater, E. (1993) *The Future of the Market. An Essay on the Regulation of Money and Nature after the Collapse of 'Actually Existing Socialism'*, London: Verso.

Altvater, E. (1994) 'Ecological and Economic Modalities of Time and Space', 76–90, in O'Connor, M. (ed.), *Is Capitalism Sustainable? Political Economy and the Politics of Ecology*, New York/London: The Guildford Press.

Altvater, E. (2005) *Das Ende des Kapitalismus wie wir ihn kennen. Eine radikale Kapitalismuskritik*, Münster: Westfälisches Dampfboot.

Altvater, E. and Brunnengräber, A. (2008) 'Mit dem Markt gegen die Klimakatastrophe? Einleitung und Überblick', 9–20, in Altvater, E. and Brunnengräber, A. (eds), *Ablasshandel gegen Klimawandel? Marktbasierte Instrumente in der globalen Klimapolitik und ihre Alternativen*, Hamburg: VSA.

Ashiagbor, D. (2005) *The European Employment Strategy*, Oxford: Oxford University Press.

Baer, P., Athanasiou, T., Kartha, S. and Kemp-Benedict, E. (2008) *The Greenhouse Development Rights Framework. The Right to Development in a Climate Constrained World*, Berlin: Heinrich Böll Foundation.

Baker, D. (2008) 'The Housing Bubble and the Financial Crisis', *Real-World Economics Review* (46): 73–81.

Baron, R. and Philibert, C. (2005) *Act Locally, Trade Globally. Emissions Trading for Climate Policy*, Paris: OECD.

Beard, T. R. and Lozada, G. A. (1999) *Economics, Entropy and the Environment. The Extraordinary Economics of Nicholas Georgescu-Roegen*, Cheltenham: Edward Elgar.

Beck, U. (1992) *Risk Society: Towards a New Modernity*. New Delhi: Sage.

Becker, J. (2002) *Akkumulation, Regulation, Territorium. Zur kritischen Rekonstruktion der französischen Regulations Theorie*, Marburg: Metropolis.

Belk, R. (1988) 'Possessions and the Extended Self', *Journal of Consumer Research* 15: 139–68.

Bertrand, H. (2002) 'The Wage-Labour Nexus', 80–6, in Boyer, R. and Saillard, Y. (eds), *Regulation Theory. The State of the Art*, London: Taylor and Francis.

Bischoff, J. (2008) 'Globale Wirtschaftskrise. Deutungsansätze und Bausteine zur theoretischen Einordnung', 27–44, in Altvater, E., Bischoff, J., Hickel,

R., Hirschel, D., Huffschmid, J. and Zinn, K. G. (eds), *Krisen Analysen*, Hamburg: VSA.

Boden, S. (2007) 'Consuming Pleasure on the Wedding Day: The Lived Experience of Being a Bride', 109–22, in Casey, E. and Martens, L. (eds), *Gender and Consumption. Domestic Cultures and the Commercialisation of Everyday Life*, Aldershot: Ashgate.

Bordo, M. D. and Eichengreen, B. (eds) (1993) *A Retrospective on the Bretton Woods System. Lessons for International Monetary Reform*, Chicago: The University of Chicago Press.

Bourdieu, P. (1984a) *Distinction: A Social Critique of Judgement and Taste*, Cambridge: Harvard University Press.

Bourdieu, P (1984b) *Homo Academicus*, Stanford: Stanford University Press.

Bourdieu, P. (1991) 'Genesis and Structure of the Religious Field', *Comparative Social Research* 13: 1–44.

Bourdieu, P. (2000) *Pascalian Meditations*, Cambridge: Polity.

Bourdieu, P. (2005) *The Social Structures of the Economy*, Cambridge: Polity.

Bourdieu, P. et al. (2002) *The Weight of the World. Social Suffering in Contemporary Society*, Cambridge: Polity.

Boyer, R. (2000) 'Is a Finance-led Growth Regime a Viable Alternative to Fordism? A Preliminary Analysis', *Economy and Society* 29 (1): 111–45.

Boyer, R. (2002) 'Perspectives on the Wage-Labour Nexus', 73–9, in Boyer, R. and Saillard, Y. (eds), *Régulation Theory. The State of the Art*, London: Taylor & Francis.

Boyer, R. and Hollingsworth. J. (1997) *Contemporary Capitalism: The Embeddedness of Institutions*, Cambridge: Cambridge University Press.

Brenner, N. (2004) 'Urban Governance and Production of New State Spaces in Western Europe, 1960–2000', *Review of International Political Economy* 11 (3): 447–88.

Broecker, W. and Kunzig, R. (2008) *Fixing Climate*, New York: Hill & Wang.

Brouns, B. and Witt, U. (2008) 'Klimaschutz als Gelddruckmaschine', 67–87, in Altvater, E. and Brunnengräber, A. (eds), *Ablasshandel gegen Klimawandel? Marktbasierte Instrumente in der globalen Klimapolitik und ihre Alternativen*, Hamburg: VSA.

Brunnengräber, A., Dietz, C., Hirschl, B., Walk, H. and Weber, M. (2008) *Das Klima neu denken. Eine sozial-ökologische Perspektive auf die lokale, nationale und internationale Klimapolitik*, Münster: Westfälisches Dampfboot.

Burkett, P. (1999) *Marx and Nature. A Red and Green Perspective*, New York: St. Martin's Press.

Burkett, P. (2005) 'Entropy in Ecological Economics: A Marxist Intervention', *Historical Materialism* 13 (1): 117–52.

Burkett, P. and Foster, J. B. (2006) 'Metabolism, Energy, and Entropy in Marx's Critique of Political Economy: Beyond the Podolinksy Myth', *Theory and Society* 35 (1): 109–56.

Chandler, A. P. (1977) *The Visible Hand: The Managerial Revolution in American Business*, Cambridge: Belknap.

Chua, B. H. (2009) 'From Small Objects to Cars: Consumption Expansion in East Asia', 101–18 in Lange, H. and Meier, L. (eds), *The New Middle Classes. Globalizing Lifestyles, Consumerism and Environmental Concern*, Dordrecht, Heidelberg, London and New York: Springer.

Cieraad, I. (2007) 'Gender at Play: Décor Differences Between Boys and Girls', 197–218, in Casey, E. and Martens, L. (eds), *Gender and Consumption. Domestic Cultures and the Commercialisation of Everyday Life*, Aldershot: Ashgate.

Clark, B. and York, R. (2005) 'Carbon Metabolism: Global Capitalism, Climate Change, and the Biospheric Rift', *Theory and Society* 34 (4): 391–428.

Clarke, A. J. (2007) 'Making Sameness: Mothering, Commerce and the Culture of Children's Birthday Parties', 79–96, in Casey, E. and Martens, L. (eds), *Gender and Consumption. Domestic Cultures and the Commercialisation of Everyday Life*, Aldershot: Ashgate.

Cleveland, C. J. (1999) 'Biophysical Economics: From Physiocracy to Ecological Economics and Industrial Ecology', 125–54, in Mayumi, K. and Gowdy, J. M. (eds), *Bioeconomics and Sustainability. Essays in Honor of Georgescu-Roegen*, Cheltenham: Edward Elgar.

Coase, R. (1960) 'The Problem of Social Cost', *Journal of Law and Economics* 3 (1): 1–44.

Committee on Climate Change (2008) *Building a Low Carbon Economy – The UK's Contribution to Tackling Climate Change*, London: TSO.

Cowen, T. (1988) 'Public Goods and Externalities: Old and New Perspectives', 1–28 in Cowen, T. (ed.), *The Theory of Market Failure: A Critical Examination*, Fairfax: George Mason University Press.

Dales, J. H. (1969) 'Land, Water and Ownership', *Canadian Journal of Economics* 1 (4): 791–804.

Daly, H. (1974) 'The Economics of the Steady State', *American Economic Review* 64 (2): 15–21.

Daly, H. (1985) 'The Circular Flow of Exchange Value and the Linear Throughput of Matter–Energy: A Case of Misplaced Concreteness', *Review of Social Economy* 43 (3): 279–97.

Dawley, A. (1976) *Class and Community: The Industrial Revolution in Lynn*, Cambridge: Harvard University Press.

De Gleria, S. (1999) 'Nicholas Gerogescu-Roegen's Approach to Economic Value: A Theory Based on Nature with Man as its Core', 82–102, in Mayumi, K. and Gowdy, J. M. (eds), *Bioeconomics and Sustainability. Essays in Honor of Georgescu-Roegen*, Cheltenham: Edward Elgar.

Deléage, J. P. (1994) 'Eco-Marxist Critique of Political Economy', 37–52, in O'Connor, M. (ed.), *Is Capitalism Sustainable? Political Economy and the Politics of Ecology*, New York/London: The Guildford Press.

Demsetz, H. (1969) 'Information and Efficiency: Another Viewpoint', *Journal of Law & Economics* 12 (1): 1–22.

Dietz, K. (2009) 'Prima Klima in den Nord-Süd-Beziehungen? Die Antinomien globaler Klimapolitik: Diskurse, Politiken und Prozesse', 183–218, in Burchardt, H. J. (ed.), *Nord-Süd-Beziehungen im Umbruch. Neue Perspektiven auf Staat und Demokratie in der Weltpolitik*, Frankfurt am Main/New York: Campus.

Dietz, K. and Wissen, M. (2009) 'Kapitalismus und "natürliche Grenzen" Eine kritische Diskussion ökomarxistischer Zugänge zur ökologischen Krise', *Prokla* 39 (3): 351–69.

Dittrich, C. (2009) 'The Changing Food Scenario and the Middle Classes in the Emerging Megacity of Hyderabad, India', 269–80, in Lange, H. and Meier, L. (eds), *The New Middle Classes. Globalizing Lifestyles, Consumerism and Environmental Concern*, Dordrecht, Heidelberg, London and New York: Springer.

Durkheim, E. (1997) *On the Division of Labour in Society*, New York: The Free Press.

Durkheim, E. (2001) *The Elementary Forms of the Religious Life*, Oxford: Oxford University Press.

Ebert, A. (2010) 'Der Clean Development Mechanism – Ein Beitrag zur nachhaltigen Entwicklung?', *Berliner Debatte Initial* 21 (1): 23–7.

EIA (2008) *International Energy Annual 2006*, Washington, DC: Energy Administration Organisation; available online at http://www.eia.doe.gov (retrieved 05 July 2010).

Eriksen, S., Klein, R., Ulsrud, K. Naess, L. and O'Brien, K. (2007) 'Climate Change Adaptation and Poverty Reduction: Key Interactions and Critical Measures', Report prepared for the Norwegian Agency for Development Cooperation (NORAD), Oslo.

Esping-Andersen, G. (1990) *The Three Worlds of Welfare Capitalism*, Cambridge: Polity Press.

EU Commission (2006) Press Release from 29.11.2006: Emissions trading: Commission Decides on First Sets of National Allocation Plans for the 2008–2012 Trading Period, avsilable at http://europa.eu/rapid/pressReleasesAction. do?reference= IP/06/1650&format= HTML&aged= 0&language= EN& guiLanguage= en, Brussels (retrieved 24 July 2010).

EU Commission (2008a) Press Release from 23.05.2008: Emissions trading: 2007 verified emissions from EU ETS businesses, available at http://europa.eu/ rapid/pressReleasesAction.do?reference= IP/08/787&format= HTML&aged= 0 &language= EN&guiLanguage= en, Brussels (retrieved 22 July 2010).

EU Commission (2008b) Memo/08/35 from 23.01.2008: Questions and Answers on the Commission's Proposal to Revise the EU Emissions Trading System, available at http://europa.eu/rapid/pressReleasesAction.do?reference= MEMO/ 08/35&format= HTML&aged= 0&language= EN&guiLanguage= en (retrieved 25 July 2010).

EU Commission (2010) Press Release from 18.05.2010: Emissions Trading: EU ETS Emissions Fall More than 11 % in 2009, available at http://europa.eu/ rapid/pressReleasesAction.do?reference= IP/10/576&format= HTML&aged= 0 &language= EN&guiLanguage= en, Brussels (retrieved 22 July 2010).

EuroMemorandum Group (2009) *Europe in Crisis: A Critique of the EU's Failure to Respond – EuroMemorandum 2009/2010*, Bremen University.

Farrell, D., Fölster, C. S. and Lund, S. (2008) 'Long-term Trends in the Global Capital Markets', *The McKinsey Quarterly, Economic Studies*, February.

Feldbauer, P., Gächter, A., Hardach, G. and Novy, A. (eds) (1995) *Industrialisierung – Entwicklungsprozesse in Afrika, Asien und Lateinamerika*, Frankfurt/Main and Vienna: Brandes und Apsel/Südwind.

Fisahn, A. (2008) 'Vollzugsdefizite im künstlichen Markt', 51–66, in Altvater, E. and Brunnengräber, A. (eds), *Ablasshandel gegen Klimawandel? Marktbasierte Instrumente in der globalen Klimapolitik und ihre Alternativen*, Hamburg: VSA.

Foreningen Civilsamfundets Klimaforum (2010) *Klimaforum09, Peoples' Climate Summit – Evaluation Report 09*, Copenhagen, available online at www. klimaforum09.org.

Foster, J. B. (1988) 'The Fetish of Fordism', *Monthly Review* 39 (10): 14–33.

Foster, J. B. (2000) *Marx's Ecology*, New York: Monthly Review Press.

Foster, J. B. (2009) *The Ecological Revolution. Making Peace with the Planet*, New York: Monthly Review Press.

Foster, J. B. and Magdoff, F. (2009) *The Great Financial Crisis*, New York: Monthly Review Press.

Friedman, A. L. (2000) 'Microregulation and Post-Fordism: Critique and Development of Regulation Theory', *New Political Economy* 5 (1): 59–76.

Fröbel, F., Heinrichs, J. and Kreye, O. (1981) *Krisen in der kapitalistischen Weltökonomie*, Reinbek: Rowohlt.

Georgescu-Roegen, N. (1971) *The Entropy Law and the Economic Process*, Cambridge and London: Harvard University Press.

Georgescu-Roegen, N. (1981) 'Energy, Matter, and Economic Valuation: Where Do We Stand?', 43–79, in Daly, H. and Umaña, A. F. (eds), *Energy, Economics, and the Environment*, Boulder: Westview Press.

Giddens, A. (2009) *The Politics of Climate Change*, Cambridge: Polity.

Gilbertson, T. and Reyes, O. (2009) 'Carbon Trading – How it Works and Why it Fails', *Critical Currents* No. 7, Uppsala.

Girouard, N., Kennedy, M. and André, C. (2006) 'Has the Rise in Debt Made Households More Vulnerable?' OECD Economics Department Working Paper, No. 535, OECD Publishing.

Global Humanitarian Forum (2009) Human Impact Report: Climate Change – The Anatomy of a Silent Crisis, Geneva.

Görg, C. (2003) *Regulation der Naturverhältnisse. Zu einer kritischen Theorie der ökologischen Krise*, Münster: Westfälisches Dampfboot.

Görg, C. and Brand, U. (2000) 'Global Environmental Politics and Competition between Nation-States: On the Regulation of Biological Diversity', *Review of International Political Economy* 7 (3): 371–98.

Görg, C. and Brand, U. (2003) 'Post-Fordist Social Relationships with Nature: The Role of NGOs and the State in Biodiversity Politics', *Rethinking Marxism* 15 (2): 263–88.

Gordon, D. M., Edwards, R. and Reich, M. (1982) *Segmented Work: Divided Workers*, New York: Cambridge University Press.

Gowan, P. (1999) *The Global Gamble: Washington's Bid for World Dominance*, London: Verso.

Gowan, P. (2004) 'US Hegemony Today', 57–76, in Foster, J. B. and McChesney, R. W. (eds), *Pox Americana. Exposing the American Empire*, London: Pluto Press.

Gramsci, A. (1971) *Selections from the Prison Notebooks*, New York: International Publishers.

Grubb, M., Vrolijk, C. and Brack, D. (1999) *The Kyoto Protocol: A Guide and Assessment*, London: Royal Institute of International Affairs.

Grubb, M., Brewer, T. L., Sato, M., Heilmayr, R. and Fazekas, D. (2009) 'Climate Policy and Industrial Competitiveness. Ten Insights from Europe on the EU Emission Trading System', Climate and Energy Paper Series 09, The German Marshall Fund of the United States.

Gupta, J. (2000) 'North–South Aspects of the Climate Change Issue: Towards a Negotiating Theory and Strategy for Developing Countries', *International Journal of Sustainable Development* 3 (2): 115–35.

Guttmann, R. (2002) 'Money and Credit in Régulation Theory', 57–63, in Boyer, R. and Saillard, Y. (eds), *Régulation Theory. The State of the Art*, London: Taylor and Francis.

Habermas, J. (1990) *Moral Consciousness and Communicative Action*, Cambridge, MA: MIT Press.

Haensgen, T. (2002) 'Das Kyoto Protokoll: Eine ökonomische Analyse unter besonderer Berücksichtigung der flexible Mechanismen', Working Paper No. 40, Bamberg Economic Research Group on Government and Growth.

Hall, A. and Midgley, J. (2004) *Social Policy for Development*, London: Sage.

Hall, P. A. and Soskice, D. W. (2001) *Varieties of Capitalism. The Institutional Foundations of Comparative Advantage*, Oxford: Oxford University Press.

Hansen, J. (2008) 'Perspective of a Climatologist', 6–15, in Woods, W. (ed.), *State of the Wild 2008–2009: A Global Portrait of Wildlife, Wildlands, and Oceans*. Washington DC: Island Press, Wildlife Conservation Society.

Hansen, J., Sato, M., Kharecha, P. et al. (2008) 'Target Atmospheric CO_2: Where Should Humanity Aim?', *The Open Atmospheric Science Journal* 2: 217–31.

Harvey, D. (1990) *The Condition of Postmodernity: An Enquiry into the Origins of Cultural Change*, Cambridge: Blackwell.

Harvey, D. (2005) *The New Imperialism*, Oxford: Oxford University Press.

Harvey, D. (2006) *Spaces for Global Capitalism*, London: Verso.

Harvey, D. (2009) *A Brief History of Neo-liberalism*, Oxford: Oxford University Press.

Haya, B. (2007) *Failed Mechanism. How the CDM is Subsidizing Hydro Developers and Harming the Kyoto Protocol*, Berkeley: International Rivers.

Heeg, S. and Oßenbrügge, J. (2002) 'State Formation and Territoriality in the European Union', *Geopolitics* 7 (3): 75–88.

Herkommer, S. (2004) *Metamorphosen der Ideologie. Zur Analyse des Neoliberalismus durch Pierre Bourdieu und aus marxistischer Perspektive*, Hamburg: VSA.

Hilferding, R. (1981) *Finance Capital. A Study of the Latest Phase of Capitalist Development*, London: Routledge & Kegan Paul.

Hirsch, F. (1976) *The Social Limits to Growth*, Cambridge: Harvard University Press.

Hirsch, J. and Roth, R. (1986) *Das neue Gesicht des Kapitalismus: vom Fordismus zum Postfordismus*, Hamburg: VSA.

Hounshell, D. A. (1984) *From the American System to Mass Production, 1800–1932: The Development of Manufacturing Technology in the United States*, Baltimore, MD: John Hopkins University Press.

Huffschmid, J. (2009) 'Europäische Perspektiven im Kampf gegen die Wirtschafts- und Finanzkrise', 105–18, in Altvater, E., Bischoff, J., Hickel, R., Hirschel, D., Huffschmid, J. and Zinn, K. G. (eds), *Krisen Analysen*, Hamburg: VSA.

Hughes, T. P. (2004) *American Genesis: A Century of Invention and Technological Enthusiasm 1870–1970*, Chicago: University of Chicago Press.

IEA (2007) *World Energy Outlook 2007*, Paris: China and India Insights.

IEA (2008) *Key World Energy Statistics*, Paris.

IEA (2009) *Key World Energy Statistics*, Paris.

The International Labour Organisation (ILO) (2009) 'Tackling the Global Jobs Crisis: Recovery through Decent Work Policies', International Labour Conference, 98th Session 2009, Report I (A), Geneva.

IPCC (1995) *IPCC Second Assessment: Climate Change 1995*, IPCC reports, WMO, UNEP: Geneva.

IPCC (Solomon, S., Qin, D., Manning, M. et al. (eds) (2007a) 'Summary for Policymakers', in *Climate Change 2007: The Physical Science Basis. Contribution of*

Working Group I to the Fourth Assessment Report of the Intergovernmental Panel on Climate Change, Cambridge: Cambridge University Press.

IPCC (2007b) *Climate Change 2007: Mitigation. Contribution of Working Group III to the Fourth Assessment Report of the Intergovernmental Panel on Climate Change*, Cambridge: Cambridge University Press.

Jachtenfuchs, M. (2001) 'The Governance Approach to European Integration', *Journal of Common Market Studies* 39 (2): 245–64.

Jackson, T. (2009) *Prosperity without Growth? The Transition to a Sustainable Economy*, London: Sustainable Development Commission.

Jessop, B. (1990) 'Regulation Theories in Retrospect and Prospect', *Economy and Society* 19 (2), 153–216.

Jessop, B. (1999) 'The Changing Governance of Welfare. Recent Trends in its Primary Functions, Scale, and Modes of Coordination', *Social Policy and Administration* 33 (4): 348–59.

Jessop, B. (2002) *The Future of the Capitalist State*, Cambridge: Polity.

Jevons, W. S. (1865) *The Coal Question. An Inquiry Concerning the Progress of the Nation, and the Probable Exhaustion of Our Coal-Mines*, London: Macmillan and Co.

Kandlikar, M. and Sagar, A. (1999) 'Climate Change Research and Analysis in India: An Integrated Assessment of a South-North Divide', *Global Environmental Change* 9 (2): 119–38.

Koch, M. (1996) 'Class and Taste. Bourdieu's Contribution to the Analysis of Social Structure and Social Space', *International Journal of Contemporary Sociology* 33 (2): 187–202.

Koch, M. (1998a) *Vom Strukturwandel einer Klassengesellschaft. Theoretische Diskussion und Empirische Analyse* (2nd edition), Münster: Westfälisches Dampfboot.

Koch, M. (1998b) *Unternehmen Transformation. Sozialstruktur und gesellschaftlicher Wandel in Chile*, Frankfurt am Main: Vervuert.

Koch, M. (1999) 'Changes in the Chilean Social Structure: Class Structure and Income Distribution between 1972 and 1994', *European Review of Latin American and Caribbean Studies/Rivista Europea de Estudios Latinoamericanos y del Caribe* 66: 3–19.

Koch, M. (2000) 'The Theory of Sociological Thought and the Research Process', 21–34, in G. C. Kinloch and R. P. Mohan (eds), *Ideology and the Social Sciences*, Westport/London: Greenwood Press.

Koch, M. (2001) 'In Search of a Class Theory of Marginality and Exclusion', *International Journal of Contemporary Sociology* 38 (2): 193–212.

Koch, M. (2003) *Arbeitsmärkte und Sozialstrukturen in Europa. Wege zum Postfordismus in den Niederlanden, Schweden, Spanien, Großbritannien und Deutschland*, Wiesbaden: Westdeutscher Verlag.

Koch, M. (2005) 'Wage Determination, Socio-Economic Regulation and the State', *European Journal of Industrial Relations* 11 (3): 327–46.

Koch, M. (2006a) *Roads to Post-Fordism. Labour Markets and Social Structures in Europe*, Aldershot: Ashgate.

Koch, M. (2006b) 'Pierre Bourdieu as a Sociologist of the Economy and Critic of "Globalisation" ', *International Journal of Contemporary Sociology* 43 (1): 71–86.

Koch, M. (2008) 'The State in European Employment Regulation', *Journal of European Integration* 30 (2): 255–272.

Koch, M. (2009) 'Klassen- und Sozialstrukturanalyse in transnationaler Dimension', 273–304, in Burchardt, H. G. (ed.), *Nord–Süd Beziehungen im Umbruch. Neue Perspektiven auf Staat und Demokratie in der Weltpolitik*, Frankfurt am Main/New York: Campus.

Koch, M. (2010) 'Klassenstrukturen in Europa: Zwischen Homogenisierung und Vertiefung', 310–35, in Thien, H. G. (ed.), *Klassen im Postfordismus*, Münster: Westfälisches Dampfboot.

Koch, M. (forthcoming) 'Poulantzas's Contribution to Class Analysis and Social Structural Analysis', in Gallas, A, Bretthauer, L., Kannankulam, J. and Stützle, I. (eds), *Reading Poulantzas*, London: Merlin Press.

Koch, M., McMillan, L. and Peper, B. (2011) 'Diversity, Standardization and the Perspective of Social Transformation', 213-22, in Koch, M., McMillan, L. and Peper, B. (eds), *Diversity, Standardization and Social Transformation. Gender, Ethnicity and Inequality in Europe*, Aldershot: Ashgate.

Krugman, P. (2009) *The Return of Depression Economics and the Crisis of 2008*, New York/London: W. W. Norton & Company.

Laeven, L. and Valencis, F. (2008) 'Systematic Banking Crises: A New Database', International Monetary Fund, Working Paper aWP/08/224, Washington, DC.

Leisink, P. and Hyman, R. (2005) 'Introduction: The Dual Evolution of Europeanization and Varieties of Governance', *European Journal of Industrial Relations* 11 (3): 277–286.

Lenin, V. I. (1947) *Selected Works*, Moscow: Foreign Language Publishing House.

Le Treut, J., Somerville, R., Cubasch, U. et al. (2007) 'Historical Overview of Climate Change', in Solomon, S. Qin, D., Manning, M. et al. (eds), *The Physical Science Basis. Contribution of Working Group I to the Fourth Assessment Report of the Intergovernmental Panel on Climate Change*, Cambridge: Cambridge University Press.

Lipietz, A. (1998) *Nach dem Ende des 'Goldenen Zeitalters'. Regulation und Transformation kapitalistischer Gesellschaften*, Berlin: Argument Verlag.

Lipsey, R. E. (1963) *Price and Quantity Trends in the Foreign Trade of the United States*, Princeton: Princeton University Press.

Lockwood, D. (1992) *Solidarity and Schism. 'The Problem of Disorder' in Durkheimian and Marxist Sociology*, Oxford: Oxford University Press.

Lohmann, L. (2006) 'Carbon Trading. A Critical Conversation on Climate Change, Privatisation and Power', Development Dialogue No. 48, Uppsala: The Dag Hammarskjöld Centre.

Lutz, B. (1989) *Der kurze Traum immerwährender Prosperität*, Frankfurt am Main/New York: Campus.

Maddison, A. (1982) *Phases of Capitalist Development*, Oxford: Oxford University Press.

Maddison, A. (1995) *Explaining the Economic Performance of Nations. Essays in Time and Space*, Aldershot: Edward Elgar.

Maddison, A. (2007) *Contours of the World Economy, 1–2030 AD*, Oxford: Oxford University Press.

Maier, C. S. (1970) 'Between Taylorism and Technocracy: European Ideologies and the Vision of Industrial Productivity in the 1920s', *Journal of Contemporary History* 5 (2): 27–61.

Marcus, A. and Segal, H. (1989) *Technology in America. A Brief History*, San Diego/New York: Harcourt Brace Jovanovich.

Marx, K. (1988) 'Letter to Ludwig Kugelmann from 11-07-1868', 67, in Marx, K. and Engels, F. (eds), *Collected Works*, Vol. 43, New York: International Publishers.

Marx, K. (1961) *Capital: A Critique of Political Economy*, Vol. 1, Moscow: Foreign Languages Publishing House.

Marx, K. (1973) *Grundrisse: Foundations of the Critique of Political Economy*, Harmondsworth: Penguin.

Marx, K. (1974) 'Theorien über den Mehrwert. Zweiter Teil', in *Marx–Engels–Werke (MEW)*, Vol. 26.2, Berlin: Dietz.

Marx, K. (2006) *Capital: A Critique of Political Economy*, Vol. 3, London: Penguin Classics.

Massarat, M. (2008) 'Eine neue Philosophie des Klimaschutzes', 199–214, in Altvater, E. and Brunnengräber, A. (eds), *Ablasshandel gegen Klimawandel? Marktbasierte Instrumente in der globalen Klimapolitik und ihre Alternativen*, Hamburg: VSA.

Mau, S. (2009) 'Who are the Globalizers? The Role of Education and Educational Elites', 65–80, in Lange, H. and Meier, L. (eds), *The New Middle Classes. Globalizing Lifestyles, Consumerism and Environmental Concern*, Dordrecht, Heidelberg, London and New York: Springer.

Mayntz, R. (2009) 'Nachhaltige Entwicklung und Governance – neue theoretische Anforderungen', 163–82 in Burchardt, H. J. (ed.), *Nord-Süd-Beziehungen im Umbruch. Neue Perspektiven auf Staat und Demokratie in der Weltpolitik*, Frankfurt am Main/New York: Campus.

Meadows, D. H., Meadows, D. L., Randers, J. and Behrens, W. (1972) *The Limits to Growth*, New York: Universe Books.

Mesner, S. and Gowdy, J. M. (1999) 'Geogescu-Roegen's Evolutionary Economics', 51–68, in Mayumi, K. and Gowdy, J. M. (eds), *Bioeconomics and Sustainability. Essays in Honor of Georgescu-Roegen*, Cheltenham: Edward Elgar.

Michaelowa, A. and Purohit, P. (2007) 'Additionality Determination of Indian CDM Projects. Can Indian CDM Project Developers Outwit the CDM Executive Board?', Discussion Paper CDM-1, Climate Strategies, London.

Miernyk, W. H. (1999) 'Economic Growth Theory and Georgescu-Roegen Paradigm', 69–81, in Mayumi, K. and Gowdy, J. M. (eds), *Bioeconomics and Sustainability. Essays in Honor of Gerogescu-Roegen*, Cheltenham: Edward Elgar.

Milanovic, B. (2002) 'True World Income Distribution, 1988 and 1993: First Calculation Based on Household Surveys Alone', *The Economic Journal* 112 (476): 51–92.

Milanovic, B. (2005) *Worlds Apart. Measuring International and Global Inequality*, New Jersey: Princeton University Press.

Minsky, H. (1982) *Can 'It' Happen Again?* New York: M. E. Sharpe.

Musgrave, P. (1975) *Direct Investment Abroad and the Multinationals: Effects on the United States Economy*, Washington, DC: Government Printing Office.

Najam, A., Huq, S. and Sokona, Y. (2003) 'Climate Negotiations beyond Kyoto: Developing Countries Concerns and Interests', *Climate Policy* 3 (3): 221–31.

Nell, E., Semmler, W. and Rezai, A. (2008) 'Wirtschaftswachstum und globale Klimaerwärmung', 169–85, in Altvater, E. and Brunnengräber, A. (eds), *Ablasshandel gegen Klimawandel? Marktbasierte Instrumente in der globalen Klimapolitik und ihre Alternativen*, Hamburg: VSA.

O'Connor, J. (1988) 'Capitalism, Nature, Socialism. A Theoretical Introduction', *Capitalism, Nature, Socialism* 1 (1): 11–38.

O'Connor, J. (1998) *Natural Causes. Essays in Ecological Marxism*, New York/London: The Guildford Press.

O'Connor, M. (1994) 'Introduction: Liberate, Accumulate – and Bust?' 1–21, in O'Connor, M. (ed.), *Is Capitalism Sustainable? Political Economy and the Politics of Ecology*, New York/London: The Guildford Press.

Osorio, V. and Cabezas, I. (1995) *Los Hijos de Pinochet*, Santiago de Chile: Editorial Planeta Chilena.

Panitch, L. and Konings, M. (2009) *American Empire and the Political Economy of Global Finance*, Basingstoke: Palgrave Macmillan.

Pigou, A. C. (1932) *The Economics of Welfare*, London: Macmillan.

Polanyi, K. (1944) *The Great Transformation. The Political and Economic Origins of Our Time*, Boston: Beacon Press.

Poulantzas, Nicos (1975) *Political Power and Social Classes*, London: Verso.

Ptak, R. (2008) 'Wie ein Markt entsteht und aus Klimamüll eine Ware wird', 35–50, in Altvater, E. and Brunnengräber, A. (eds), *Ablasshandel gegen Klimawandel? Marktbasierte Instrumente in der globalen Klimapolitik und ihre Alternativen*, Hamburg: VSA.

Rae, J. B. (1969) *Henry Ford*, New Jersey: Prentice-Hall.

Reusswig, F. and Isensee, A. (2009) 'Rising Capitalism, Emerging Middle-Classes and Environmental Perspectives', 119–42, in Lange, H. and Meier, L. (eds), *The New Middle Classes. Globalizing Lifestyles, Consumerism and Environmental Concern*, Dordrecht, Heidelberg, London and New York: Springer.

Revelli, M. (1997) '*Vom "Fordismus" zum "Toyotismus". Das kapitalistische Wirtschafts- und Sozialmodell im Übergang*, Hamburg: VSA.

Rhodes, M. (2000) 'Restructuring the British Welfare State: Between Domestic Constraints and Global Imperatives', in Scharpf, F. W. and Schmidt, V. A. (eds), *Welfare and Work in the Open Economy*, Vol. 2, Oxford: Oxford University Press.

Richards, M. (2001) *A Review of the Effectiveness of Developing Country Participation in the Climate Change Convention Negotiations*, London: Overseas Development Institute.

Roberts, J. T. and Parks, B. C. (2006) *A Climate of Injustice. Global Inequality, North-South Politics, and Climate Policy*, Cambridge, MA: The MIT Press.

Robinson, R. V. and Briggs, C. M. (1991) 'The Rise of Factories in Nineteenth-Century Indianapolis', *American Journal of Sociology* 97 (3): 622–56.

Sakowsky, D. (1992) 'Die Wirtschaftspolitik der Regierung Thatcher', Wirtschaftswissenschaftliche Monographien 10, Göttingen University.

Scharpf, F. W. (1993) 'Coordination in Hierarchies and Networks', 125–65, in Scharpf, F. W. (ed.), *Games in Hierarchies and Networks. Analytical and Empirical Approaches to the Study of Governance Institutions*, Frankfurt am Main/New York: Campus.

Scheer, H. (2007) *Energy Autonomy. The Economic, Social and Technological Case for Renewable Energy*, London: Earthscan.

Schneider, L. (2007) 'Is the CDM Fulfilling its Environmental and Sustainable Development Objectives? An Evaluation of the CDM and Options for Improvement', *Report prepared for the WWF*, Berlin: Öko-Institut.

Schreurs, M. A. (2008) 'Was uns die bisherigen Erfahrungen lehren', 21–34, in Altvater, E. and Brunnengräber, A. (eds), *Ablasshandel gegen Klimawandel?*

Marktbasierte Instrumente in der globalen Klimapolitik und ihre Alternativen, Hamburg: VSA.

Schwartzman, D. (2008) 'The Limits to Entropy: Continuing Misuse of Thermodynamics in Environmental and Marxist Theory', *Science & Society* 72 (1): 43–62.

Sen, A. (1999) *Development as Freedom*, Oxford: Oxford University Press.

Shiller, R. (2006) *Irrational Exuberance* (2nd edition), Princeton: Princeton University Press.

Simms, A. (2005) *Ecological Debt. The Health oft the Planet and the Wealth of Nations*, London: Pluto Press.

Sorensen, C. E. (1956) *My Forty Years with Ford*, New York: W.W. Norton.

Stern, N. (2007) *The Economics of Climate Change: The Stern Review*, Cambridge: Cambridge University Press.

Stern, N. (2009) *A Blueprint for a Safer Planet. How to Manage Climate Change and Create a New Era of Progress and Prosperity*, London: The Bodley Head.

Stiglitz, J. (2010) *Freefall. Free Markets and the Sinking of the Global Economy*, London: Penguin Books.

Stiglitz, J., Ocampo, A., Spiegel, S. French-Davis, R. and Nayyar, D. (2006) *Stability with Growth*, Oxford: Oxford University Press.

Stockhammer, E. (2008) 'Some Stylized Facts on the Finance-Dominated Accumulation Regime', *Competition and Change* 12 (2): 184–202.

Taylor, F. W. (1947) *Scientific Management*, New York: Harper & Row.

Tickel, A. and Peck, J. (1995) 'Social Regulation after Fordism: Regulation Theory, Neo-Liberalism and the Global–Local Nexus', *Economy and Society* 24 (3): 357–86.

Trexler, M.C., Broekhoff, D.J. and Kosloff, L.H., 2006 'A statistically-driven approach to offset-based GHG additionality determinations: What can we learn?' Sustainable Development Law & Policy, 6 (2): 30–40.

UNCTAD (2007) *World Investment Report*, New York/Geneva: United Nations Publication.

UNCTAD (2009) *World Investment Report*, New York/Geneva: United Nations Publication.

UNDP (2007) *Human Development Report 2007/2008. Fighting Climate Change: Human Solidarity in a Divided World*, Basingstoke: Palgrave Macmillan.

UNFCCC (2001), 'Report of the global environment facility to the conference, note by the secretariat', FCCC/CP/2001/8, United Nations Framework Convention on Climate Change, United Nations Office at Geneva, Geneva, Switzerland.

Valdés, J. G. (1989) *La Escuela de Chicago: Operación Chile*, Buenos Aires: Editorial Zeta.

Vanek, J. (1963) *The Natural Resource Content of United States Foreign Trade 1870–1955*, Cambridge, MA: MIT Press.

Voß, J. P. (2007) 'Designs of Governance. Development of Policy Instruments and Dynamics in Governance', PhD thesis at the University of Twente.

Wara, M. (2007) 'Is the Global Carbon Market Working?' *Nature* 445: 595–6.

Weber, M. (1978) *Economy and Society. An Outline of Interpretive Sociology*, Berkeley: University of California Press.

Weber, M. (1986) 'Die protestantische Ethik und der Geist des Kapitalismus', in *Gesammelte Aufsätze zur Religionssoziologie*, Vol. 1, Tübingen: Mohr.

Wei, Y.-M., Liu, C., Fan, Y. and Wu, G. (2007) 'The Impact of Lifestyle on Energy Use and CO_2 Emission: An Empirical Analysis of China's Residents', *Energy Policy* 35 (1): 247–57.

Williams, K., Haslam, C. and Williams, J. (1992) 'Ford versus "Fordism": The Beginning of Mass Production', *Work, Employment and Society* 6 (4): 517–555.

Witt, U. and Moritz, F. (2008) 'CDM – saubere Entwicklung und dubiose Geschäfte', 88–105, in Altvater, E. and Brunnengräber, A. (eds), *Ablasshandel gegen Klimawandel? Marktbasierte Instrumente in der globalen Klimapolitik und ihre Alternativen*, Hamburg: VSA.

World Bank (2007) *Global Economic Prospects: Managing the Next Wave of Globalization*, Washington, DC: World Bank Publications.

World Bank (2009) *States and Trends of the Carbon Market*, Washington, DC: World Bank Publications.

Wright, G. (1990) 'The Origins of American Industrial Success, 1879–1940', *American Economic Review* 80 (4): 651–68.

WWF (2006) *Living Planet Report 2006*, Gland, London and Oakland.

Xu, L. (2007) 'The Changing Lives and Values of Chinese Rural Youth in the 1980s and 1990s', *The Chinese Historical Review* 14 (1): 59–96.

Zhang, Y., Deng J., Majumdar, S. and Zheng, B. (2009) 'Golfing in China', 143–58, in Lange, H. and Meier, L. (eds), *The New Middle Classes. Globalizing Lifestyles, Consumerism and Environmental Concern*, Dordrecht, Heidelberg, London and New York: Springer.

Ziltener, P. (2000) 'Die Veränderung von Staatlichkeit in Europa – regulations- und staatstheoretische Überlegungen', 73–101, in Bieling, H. J. and Steinhilber, J. (eds), *Die Konfiguration Europas: Dimensionen einer kritischen Integrationstheorie*, Münster: Westfälisches Dampfboot.

Index